FOSTERING INNOVATIONS FOR SUSTAINABILITY WITH

MECHANISMS FROM COMMUNITIES OF INNOVATORS

By

JANE ANN TALKINGTON

Bachelor of Science in Marketing
Oklahoma State University
Stillwater, Oklahoma
1990

Master of Business Administration
University of Tulsa
Tulsa, Oklahoma
2000

Submitted to the Faculty of the
Graduate College of the
Oklahoma State University
in partial fulfillment of
the requirements for
the Degree of
DOCTOR OF PHILOSOPHY
July 2016

FOSTERING INNOVATIONS FOR SUSTAINABILITY WITH

MECHANISMS FROM COMMUNITIES OF INNOVATORS

Dissertation Approved:

_____

Dr. Craig Watters, Dissertation Advisor

_____

Dr. Khaled Mansy

_____

Dr. Arthur Stoecker

_____

Dr. Cosette Armstrong

ACKNOWLEDGEMENTS

I wish to thank my support networks as well as the countless people who have crossed my path who, perhaps unwittingly, contributed inspiration to my endeavors. First, thank you to my mother Patricia Jackson, the one person who has faithfully and genuinely loved me and supported me since my embryonic stages—literally and intellectually. She was an inspiration to many in my extended family—fathers, brothers, sisters, and children—who were vital through this doctoral quest. Second, my gratitude goes to Paul Hawken who so generously shared his insights through his many fabulous books and his friendship with me; you are a beacon of hope and an example of rare intellect and uncommon kindness. You inspire me. Many people crossed my path perhaps briefly, but with great impact: Sim Van der Ryn, Patrice Pike, Ray Anderson, Dr. Will Focht, Dr. Donald French, Dr. Amanda Harrist, Dr. Melanie Page, President David Schmidly, President Hargis and Mrs. Anne Hargis, President John Bardo, Dr. Harold Glasser, Michael Haskins (a kind stranger with editing skills) and my gracious, faithful friend and foe to the bitter end, President Bob of Van Kirk University. A special thanks also goes to the Wake Up & Dream Ecovillage Plan keynote speakers who graciously invested their insights with those who dared dream of the university embracing sustainability deeply in their housing models: Charles Durrett, Rick Darnaby, Daniel Greenberg, Nancy Gift, Bill Reed, Tom Kopf, Tony Layne, Fiona Cousins, and Christopher Mare. You are a dream team. And Jeff Murphy, you are my dream student who symbolizes why I write, research, and teach. I would not *see* the world as I do without all of you.

To my committee members past and present who, by virtue of endurance and interdisciplinary tolerance, I extend sincere gratitude: Dr. Craig Watters, Dr. Art Stoecker, Dr. Khaled Mansy, Dr. Cossette Armstrong, Dr. Ken Kiser, and Dr. John Mowen. To the one person who initially encouraged me to pursue the doctoral degree and who tirelessly coached me through every valley, Dr. Nizam Najd, I am absolutely positive that I would not have persevered without you in my corner. And lastly, thank you to Sharla Helton M.D., a brilliant, amazing, and courageous person, who taught me the absolute necessity of accurate and ethical research. For you, I cross the finish line. You gave me a glimpse into the 'real world' that created an entirely new perspective on peer-review articles. You are the embodiment of sustainability by example of sacrificing your energies for future generations and for people living now all over the world that you will never know. In the strangest of ways, studying sustainability for a decade prepared me to be of help in your quest. Thank you all—*each and every one of you*—for believing in me.

Acknowledgements reflect the views of the author and are not endorsed by committee members or Oklahoma State University

DEDICATION

To the sages of the twenty-first century: after years of chasing feral thoughts I captured a few with ink just to inspire you, the friends I will never meet and the university leaders for whom helping others invent sustainability is paramount. To my mother of the twentieth century: thank you for constantly telling me I could accomplish anything I set my mind to do.

Acknowledgements reflect the views of the author and are not endorsed by committee members or Oklahoma State University

Name: JANE TALKINGTON

Date of Degree: JULY 2016

Title of Study: FOSTERING INNOVATIONS FOR SUSTAINABILITY WITH

MECHANISMS FROM COMMUNITIES OF INNOVATORS

Major Field: ENVIRONMENTAL SCIENCE

Abstract:

Universities are positioned to play a pivotal role in educating a populous with the ability to operate the planet in accordance with sustainability principles and practices. Within their responsibility is the task to prepare citizens to operate future civilizations. This is in addition to the previous missions of educating for inquiry, scholarship, or career development. Increasingly, universities are recognizing that meaningful, deep learning happens outside the curriculum and in social settings, some of which are designed specifically to foster innovation and entrepreneurship through innovation ecosystems. The result has been the creation of live-learn dormitories, living labs, innovation campuses, corporate co-location buildings, and innovation districts. In parallel, global societies are also gradually realizing that any peaceful and prosperous society in the future will require citizens who cultivate the sustainability ethic necessary to innovate solutions to the environmental challenges and social equity problems that threaten the continuation of humanity.

The challenge facing higher education then becomes the question of what new role the university can envision for itself that propels them to create or participate in the creation of a physical place with an entrepreneurial culture and a social support network capable of fostering the development of system-thinking innovators of sustainable solutions. By applying the process tracing method within a historical research methodology, this dissertation reviews past *communities of innovators* as found in *intentional communities* and *places of innovation* to identify the mechanisms used to approach their goals. The intent of the research is to understand more clearly how others in the past approached what was perceived as unimaginable so that this generation can build the confidence and courage to tackle humanity's goals that now seem unachievable. This dissertation finds that those mechanisms are open source, iterative processes, and proximity. This dissertation concludes with implementation scenarios of how these mechanisms can be integrated into strategy by higher education institutions that are striving to create the environments capable of fueling the development of *sustainability-oriented innovations*.

TABLE OF CONTENTS

LIST OF TABLES

LIST OF FIGURES

CHAPTER I

INTRODUCTION

Many people have pondered what the role of higher education could be in the creation

toward a sustainable society. It is a challenging matter. As the poet Rainer Maria Rilke

penned as advice to a friend over 100 years ago:

> Be patient toward all that is unsolved in your heart and try to love the questions
> themselves, like locked rooms and like books that are now written in a very
> foreign tongue. Do not now seek the answers, which cannot be given you because
> you would not be able to live them. And the point is, to live everything. Live the
> questions now. Perhaps you will then gradually, without noticing it, live along
> some distant day into the answer. (Rilke & Burnham, 1993)

Based on the number of people that have pondered such a role, it would seem that people

do indeed 'love the questions' around innovating sustainability from within the higher

education system, but as institutions we have yet to 'live the questions' on a day-to-day

basis. As this poet notes, it is only through patiently living the question that the answers

will gradually and subtly manifest.

There is a reason people love these questions about sustainability but have not quite

managed to live them: "Sustainability is ultimately utopian in nature; it is a high ideal to

strive for and one toward which we must progress, but nonetheless an idea that, in reality,

1

is probably not completely knowable or achievable" (J. P. Lockyer, 2007). The utopian nature of sustainability makes 'loving the questions' easy; people like to imagine utopian futures. Utopianism itself "originates in the human propensity to dream," and is even referred to grandly as "social dreaming" (Sargent, 1994). Utopianism is psychologically appealing, and insofar as sustainability partakes of utopianism, the dream of sustainability is easy to find mesmerizing. However, if a dream is considered unachievable or if every experiment is expected to fail, then it is easy to see why sustainability is an intriguing topic to discuss on university campuses, yet an elusive one to genuinely analyze, understand, or even measure. It is easy to see why people love asking the questions about sustainability but are reluctant to live them. If people do not think sustainability can be accomplished, if people think sustainability is doomed to fail, then putting those theoretical questions into practice becomes a low priority.

This dissertation investigates how utopian visions and innovative thinking have been put into successful, if limited, practice in the past so that we might understand how they can be fused and put into purpose-driven, widespread practice in the future. Specifically, this dissertation investigates *communities of innovators* in order to gather the lessons that they have learned and determine how those lessons can inform higher education institutions in their quest to provide environments capable of fostering the types of experimentation that lead *to sustainability-oriented innovations*. This dissertation focuses on implementation within higher education institutions because, as David Orr affirms, "No institutions in modern society are better equipped to catalyze the necessary transition to a sustainable world than colleges and universities. They have access to the leaders of tomorrow and the leaders of today. What they do matters to the wider public" (David W Orr, 2012).

The investigation begins with the two most recent forms of communities that innovate: the ecovillage settlement and the innovation district development. A *community of innovators* is just exactly that: a group of people who spend significant time together in a physical place for the purpose of bettering, or innovating, a product or process. The investigation then applies historical methodology to trace the genealogy of the ecovillage through the evolution of *intentional communities* over the past 400 years. Next, the investigation traces the genealogy of today's emerging innovation districts through other forms in the built environment, or *places of innovation*, over the past 100 years. These *communities of innovators*, past and present, reveal insights into how higher education institutions can stake their own role in the *knowledge economy* by defining their participation in fostering *sustainability-oriented innovations*. Finally, analysis of the findings results in the development an emerging theory about the mechanisms people have used to approach innovation.

Background of the Problem

There is a duality to the problem of creating a sustainable world: (1) what exactly needs to happen? And (2) whose job is it to prepare and lead society toward this ideal? Communities of innovators have left a rich history about how they approached their stated purposes and that history can be leveraged to create even more innovative solutions in the future. However, that history has not yet been adequately leveraged through investigations for that purpose. Higher education institutions—our largest institutions in charge of developing critical thinking skills and leaders—are a likely point of leverage to innovate an approach to creating solutions that will help shift the world toward sustainability (Meadows, 1999). Higher education houses the high ideals, the systems, and the technical knowledge needed for the task as well as having the potential to create an environment for collaboration that can be impactful to future generations (Ceulemans & De Prins, 2010; Cortese, 2003; Ferrer-Balas et al., 2010; Gadotti, 2010). Because of their unique position of leverage, these institutions ought to take responsibility for preparing and leading society toward the sustainability ideal. Yet the consensus by sustainability in scholars in higher education is that universities have "largely failed in the ethical obligation to prepare students to face the sustainability challenges of the common decades" (Vincent, Roberts, & Mulkey, 2015). Moreover, this responsibility is becoming increasingly obvious that it is not an option. David Orr candidly states with a sobering directness, "we are still educating the young as if there were no planetary emergency" (David W. Orr, 2004).

To be blunt, for the human species to survive the ecological decline, it will be necessary to invent an industrialized economy and way of life that does not continue to put our life support systems in peril (Wang et al., 2013). Our seemingly abundant natural resources and lack of timely economic and ecological signals have insulated us from the realization that the global civilization is operating in an unsustainable mode (Cortese, 2003). Since "we are the first generation capable of determining the habitability of the planet for humans and other species," we have a responsibility to preserve that habitability (Cortese, 2003). On that point, the Stockholm Resilience Centre released the Planetary Boundary Theory in 2009 that specified the safe operating zones of nine ecological systems that must be stable in order to have a biosphere capable of supporting human life. Of the nine planetary boundaries identified, three have already been destabilized; those are climate change, biodiversity loss, and removal of anthropogenic nitrogen from the atmosphere (Rockström, 2009). The six remaining boundaries, ocean acidification, land surface converted to cropland, freshwater availability, ozone depletion, atmospheric aerosols, and chemical pollution, are operating near the thresholds of instability. These nine areas of absolute necessity are prone to instability due to the business practices and consumer demand of the industrialized global economies that evolved over the past two centuries when there was little or no consideration for the requirements of sustainable equilibrium.

Obviously, destabilized ecological systems are unsustainable and will either collapse or recover based on the resource management approach and natural restoration ability. Although ecological destabilization is often understood as a consequence of rapid human population growth, determining a sustainable population for Earth is less about the overall number of people than the impact of those people. People impact Earth in a variety of ways, and impact should be understood to include the use of non-renewable resources, restorative time frames, and permanent damage or loss. As Mahatma Gandhi poignantly says, "Earth provides enough to satisfy every man's needs, but not every man's greed." Similarly, Buckminster Fuller, a renowned twentieth century systems thinker, made a bold statement at the end of his long career that it was "now highly feasible to take care of everybody on Earth at a higher standard of living than any have ever known" (Fuller, 1982). Perhaps the only gap of consequence is the political will to commit to developing those necessary innovations to propel the global society to satisfying 'every man's needs' while also restoring ecological stability.

The reasons higher education institutions have thus far lacked the political will to commit to sustainability, in spite of both the present ecological crisis and higher education's strong position of leverage, are complex. Higher education itself is in turmoil due to the scaling pressures of rising costs and new sources of competition that make predictable revenue questionable (Knoedler, 2015). Any industry under financial pressures in the present would have difficulty generating the foresight to treat sustainability as the paramount issue of the century (Pfeifer & Sutton, 2000). Research shows the pressing needs of today distract from using political will to focus on the long term over the short term and the collective good over the individual good (Pfeifer & Sutton, 2000). In addition, research funding for universities that was historically provided from the state and federal governments has been steadily replaced by private research dollars from corporations and individuals that bring their own research agendas that tend to benefit their immediate interests.

Most industries under this market situation have the potential to fail; higher education institutions, though they are more entrenched in the social and economic landscape, are not immune to financial collapse (Christensen & Eyring, 2011; Ehrenberg, 2012). Every generation of the last century has been materially better off than the previous generation through hard work and education. But in this century that potential is no longer guaranteed to everyone earning a college degree (G. Wagner, 1976). Young adults are increasingly considering other paths of securing a vocation that do not entail incurring a significant debt for an advanced education (J. Selingo, 2013). Higher education institutions are embracing the delivery system of massive online open courses or MOOC in an effort to expand their reach into global markets at a lower operating cost (Pappano, 2012). Massachusetts Institute of Technology launched open courseware online for free in 2012 and by March 2016, MIT had made 2,340 courses available that boasted two million visitors per month (MIT, 2016). Corporations are also increasingly concerned about the quality of the education in their recent college hires and they are engaging in new ways to evaluate the job readiness of a potential employee through nanodegrees and badges from private online education providers (Gomez, 2014; E. Porter, 2014; Sanchez-Gordon & Lujan-Mora, 2015). These trends are contributing to the gradual displacement of higher education institutions and making them more prone to financial difficulties, a position that makes many of them understandably risk-averse.

To respond to the present economic paradigm, academic fields have become highly specialized disciplines that operate as separate types of expertise rather than as a holistically organized knowledge system; this separatist approach is not conducive to a systems thinking approach suitable for sustainability. Yet, both expertise and systems thinking are needed because "designing a sustainable human future requires a paradigm shift toward a systemic perspective emphasizing collaboration and cooperation" (Cortese, 2003). The field of interdisciplinary scholarship remains a minority in the education landscape because its complex nature is fraught with challenges. Not only must interdisciplinary scholars mine multiple academic fields for theories and literature for applicable solutions for real world problems, but they must also weave this multi-disciplinary knowledge together in coherent narratives. The synthesis of divergent approaches is the elusive goal of interdisciplinary studies (Newell, 2001). This synthesized contribution is then faced with the conundrum of disseminating the knowledge back to the most appropriate academic fields, journals, or industries (Thompson Klein, 2004). The issue tackled in this dissertation – and sustainability in general – is less about a singular gap in knowledge and more about the numerous gaps between synthesized knowledge, historical insights, and holistic implementation that would foster *sustainability-oriented innovation* (Hulme, 2014). The *knowing-doing* gap refers to situations such as understanding the mechanisms used in the past and applying them to improve the future (Pfeifer & Sutton, 2000).

At present, universities educate students for existing jobs within a linear economy model that operates unsustainably (Fullan, 2006). Restructuring around sustainability means

8

educating students for jobs in a society operating in an economy based on a sustainable equilibrium model; neither those jobs nor that society even exists in substantial ways at present. President Franklin D. Roosevelt said, "we cannot always build the future for our youth, but we can build our youth for the future." While the future is unknown, there are trends that indicate how it is unfolding, such as a greening of curriculum, industries hiring sustainability-savvy system thinkers, and novel collaborative arrangements between higher education, industry and government (Fullan, 2006; Hart, Fox, Ede, & Korstad, 2015). 'Building our youth for the future' means the youth will need to be innovators of sustainability and implementers as well as advocates of the future sustainability society.

Literature Review

The investigation of the research questions focused on *communities of innovators* who had historically crafted creative and effective responses to their situations. The social response to new pressures and new possibilities is found in the history of *intentional communities*. The economic response is found in the history of *places of innovation*. Both groups have sub-types that reflect ecological responses as well. All were driven by the idea of progress, which is considered key to perpetuating a society that strives for social improvement (Sargent, 1994).

The investigation of utopian thinking, as it manifested in *intentional communities*, covers experimental communal settlements, communes, ecovillages, academic ecovillages, and innovation community prototypes. As an academic topic of inquiry, *intentional communities* have been seriously under-theorized for two reasons. First, studies tend to focus on specific communities without drawing general conclusions that could be applied to other facets of society (Sargent, 2012). Second, *intentional communities* require a complex multidisciplinary approach and most scholars in this field approach the subject from a specialty within a discipline (Sargent, 2012). William Schafer of Berea College, a social theory scholar, lamented an even deeper loss had occurred through the "scanty treatment of American history" regarding the contributions from the communitarian theorists; he considered them to be the uncelebrated vanguards of the idea of human progress and perfectibility (Schafer, 1978).

The investigation of innovative thinking, as it manifested in *places of innovation*, covers industrial districts, clusters, research parks, innovation districts, and universities. Historically, innovation has been pursued to improve the human condition while also creating personal financial gains and contributing to broader economic development. As complexity and specialization increased in the past century, the need evolved for a dedicated environment or intentionally designed place to foster innovation and industrial development on a collaborative basis. As an academic research topic, *places of innovation* have garnered interest for over 100 years. But over the past forty years such venues received extensive attention once Silicon Valley proved that ventures between industry and higher education were mutually beneficial and profitable. Also, once the Italian historical research contributions on industrial districts were translated into English in 1989, Michael Porter of Harvard initiated research on the cluster concept.

Utopianism requires innovative thinking whereas innovation is the tangible manifestation of the utopian belief that things could be imagined as better. The investigations of relevant literature and evidence were conducted under the premise that insights about past utopian applications and place-based innovation venues could be used to direct the strategy of higher education pursuits to foster innovations necessary to bring forth a sustainable world.

Problem Statement

Universities are faced with a challenging transition phase brought on by disruptions that are cultural, financial, and demographic in nature (J. J. Selingo, 2013). Organizational learning is considered to be an imperative for institutional survival, particularly in uncertain or highly competitive environments (Popper and Lipschitz, 2000). Of these many disruptions, increased competition from for-profit universities, online learning providers, and massive open online courses present tremendous threats to any institution with massive investment tied to place (Knoedler, 2015). Place, with all its overhead and geographic limitations, is also a powerful attractor and source of economic prosperity (R. Florida, 2014). In fact, long-term prosperity as a residential campus is more dependent than ever on the quality of a location in the face of competition from online education (Selingo, 2014).

In addition to pressures on the business model of higher education, there is the question of the true progress of infusing sustainability curriculum (Cullingford & Blewitt, 2013; Sydow, 2012; Wang et al., 2013). Only slivers of various disciplines have developed sustainability advocates. Overall, higher education institutions have not evolved into the leaders of the sustainability imperative that the global society needs to create the paradigm shift away from unsustainable operations (David W. Orr, 2004). Architect William McDonough makes the point that being less bad is not the same as being good any more than slowing down while driving south is not the same as driving north (Braungart, 2000).

There is benefit to higher education institutions establishing relevancy through a sustainability orientation that both perpetuates their business model and serves humanity. Society is clearly showing demand for innovations around sustainability (Kiron, Kruschwitz, Reeves, & Goh, 2013). If higher education institutions elect to position themselves in leadership roles in support of global sustainability, they not only need to amass the technical content and ethical commitment, but also need to understand the strategic mechanisms that could foster this specific kind of innovating thinking. *Communities of innovators* of the past and present provide valuable lessons learned that higher education could use to shape their institutional response to participate in innovating solutions that advance sustainability.

The problem of higher education demonstrating leadership around *sustainability-oriented innovation* is hampered by three forces that impede visionary action: unawareness about the value of past utopian and innovative thinking, a lack of understanding that the future of a viable planet requires adopting a sustainability imperative, and submission to relentless pressures to maintain current business-as-usual operations despite rapidly changing global needs. It would be inaccurate to assume that people in higher education should realize the sustainability imperatives facing humanity, but those working in higher education are just a subset of the general population that is know not to show a deep understanding of the complex environmental crises underway (Cortese, 2003).

Purpose of the Study

The purpose of the study was two-fold: first, to gather relevant evidence of how problems were approached by *communities of innovators* and, second, to use that evidence to inform innovative and entrepreneurial universities that are ready and willing to create innovations for sustainability. The scope of the work needed is daunting, but entrepreneurial universities can find solace and draw resolve from Paul Hawken's positive response to the dilemma of unsustainability: "The great thing about the dilemma we're in is that we get to reimagine every single thing we do. In other words, there isn't one single thing that we make that doesn't require a complete remake. And so there are two ways of looking at that. One is like: Oh my gosh, what a big burden. The other way to look at it, which is the way I prefer, is: What a great time to be born! What a great time to be alive! Because this generation gets to essentially completely change this world." (Z. Goldsmith, 2007). This is the kind of attitude that can implement the mechanisms identified in this dissertation and achieve tremendous progress.

The theory generated by this evidence, derived from historical analysis, suggests there are specific mechanisms that can be replicated by a university to create the conditions that will foster *sustainability-oriented innovations*. While inventions often happen by accident, creating a culture of innovation for sustainability does not; it is intentional. The purpose of this study is to provide data-driven, qualitative research to show why a university can succeed in making sustainable impacts when it applies mechanisms from utopian thinking and innovative thinking.

Importance of the Study

By 2050, the global population is predicted to reach 9.7 billion people (Nations, 2015). Managing the global resources for that many people will require applying sustainable practices so that civilization can operate within ecological limits. How many of those innovations will be developed in higher education over the coming decades is a decision for university leaders to make now.

Debra Rowe and Aurora Windslade observed that "our educational systems have so far not succeeded in teaching the skills, knowledge, and attitudes necessary to create sustainable society" (J. Martin, 2012). Some maintain the shift to an ecologically-sound society begins with green universities (Wang et al., 2013). David Orr, a sustainability thought leader in higher education, states that "the question is not whether colleges and universities could help catalyze the transition to a sustainable society, but whether they have the vision and courage do to so" (Eagan & Orr, 1992). Committing a university to an ecological twenty-first century strategy is a bold proposition; it requires intellect, courage, and wisdom. Recall sustainability is a high ideal that may not be achievable (J. P. Lockyer, 2007). But as legendary hockey player Wayne Gretzky says, "you miss 100% of the shots you don't take"—which hints that it takes courage and confidence to execute the accumulated skills.

There are many ways a university can engage in innovation. Should a university decide fostering innovations for sustainability is best implemented through participating in an innovation district, then the collaborative demands and financial commitments need to be well understood in advance. An innovation district build-out can easily span a decade and require $2 billion of investment as in the case of the Cortex Community in St. Louis (Marks, 2012). Orchestrating the social capital networks and financial partnerships are also monumental and complex tasks (Morisson, 2015). Even for a smaller or mid-sized city, an innovation district effort has the same basic complexity as the multi-million dollar projects (Glenn, 2016). Organizing a university-led *innovation ecosystem* on an existing campus is also a complex undertaking, but perhaps with less capital-intensive requirements because many of the buildings are already in place. These repositions efforts still requires forging partnerships with industry and asset mapping the expertise among the stakeholders to accomplish collaborative capacity building ("Mini Town Hall Innovation Discussions," 2015).

This study provides a unique glimpse into the mechanisms *communities of innovators* have used to approach their goals. It provides findings that can inform higher education leaders in their pursuit of fostering innovations for sustainability in: the classroom, various initiatives, university communities, research parks, and other collaborative *places of innovation*. It is actually vital to all parties and stakeholders that higher education institutions actively and intentionally engage in creating our ideal world because 'what they do matters to the wider public' (David W Orr, 2012).

Primary Research Question

The research question driving this investigation is: "What are the mechanisms historically used by *communities of innovators*, as identified in *intentional communities* and in *places of innovation*, that were used to approach their goals?" The inquiry was then tailored for the university setting. In the university context, the question driving the investigation is: "How can these mechanisms be applied to the environments created by higher education institutions so they can successfully fuel innovations that advance sustainability?"

Asking these questions helps identify the mechanisms that provide a platform for exploratory discourse into why underpinning those mechanisms with values might be fundamental to achieving sustainability. The intent of the research is to understand more clearly how others in the past approached what was perceived as unimaginable, so that this generation can build the confidence and the courage to tackle humanity's sustainability goals that now seem unachievable.

Research Design

This research was undertaken from the perspective of an environmental scientist who was molded by professional experiences in business sustainability, green building, and higher education. These fields have already reflected a demand for *sustainability-oriented innovations* and experienced tremendous changes and growth due to the contributions from their own industry advocates and innovators. History scholar E.H. Carr called on historians to create a plurality or multiplicity of causes (Sreedharan, 2007). The multiplicity of causes is the many environmental challenges and the increasing global political will to respond to those challenges; this creates the impetus for innovating solutions that move society toward ecological stability and ultimately broader sustainability. An environmental scientist has the vantage point to recognize those 'multiplicity of causes' while the professional experiences provide the validation of the response to those causes in the form of *sustainability-oriented innovation* demand.

To analyze the research for this dissertation, a qualitative methodology was chosen to interpret the data. An investigation was conducted on *communities of innovators* in the present and in the past; this included a review of both *intentional communities* and *places of innovation*. The historical analysis was guided by a technique borrowed from comparative politics in Political Science, referred to as process tracing, through which three key themes were identified in the operation of the *communities of innovators*.

Recent philosophical developments in historical analysis allows the same methodological approach used on the past to be applied to the future (Staley, 2002; Wagar, 1998). Practitioners refer to this as "scenario planning with environmental scanning" and historians who find relevant parallels with their methods now infuse scenario planning with historical analysis. What ties these two approaches together - historical analysis and scenario planning – is the interdisciplinary synthesis that uses imagination and context to create meaning from evidence in the past, in order to evaluate the present, and then forecast implications for possible trajectories in the future. Utopian scholar Gregory Claeys implores us to remember the value of history in the creation of our future ideal "The old ideal worlds can lend us hope, inspiration, a sense of what to aspire for as well as what to avoid" (Claeys, 2011).

The data were collected through process tracing techniques used to analyze evidence in *communities of innovators* identified in *intentional communities* and in *places of innovation*. In all, twelve types of *communities of innovators* were considered: *intentional communities* (also known as experimental communities), communes, ecovillages, academic ecovillages, cohousing, innovation community prototypes, industrial districts, clusters, research parks, innovation campuses, innovation districts, and universities. Any one of these types would have provided robust insights into its own specific role around innovation from within a community, but the broader view of all twelve types allows a more comprehensive review capable of revealing universal themes not evident within just a single typology.

An analogy useful for appreciating the research design would be spending ten years visiting communities in all 50 states in the USA with the goal of identifying a unified description of approaches used by American citizens. Considering the United States has regional personalities, an arbitrary geographic sample set of 50 states could be reorganized and simplified into smaller groups instead of keeping the sample size at 50. Even by just visiting a few states in the South, one could create an accurate cultural appraisal without visiting all of the southern states. A study of the nation's past and expansion doctrines might also lend credence into the historical significance found in the current sample. And so it is with the analysis of the *communities of innovators* that are very different in type and individually unique, yet evolutionarily related in a way that is reflected in some common mechanisms.

The investigation for the dissertation gathered evidence from academic publications, public documents, non-governmental organizations, historical books, and trade industry reports. Academic fields scoured included urban planning, architecture, utopian studies, communal studies, social movements, economic geography, business, history, design, community development, higher education, entrepreneurship, sustainability, and innovation. The majority of the data are from evidence from the past two decades but some points were discovered in all but forgotten texts teeming with historical context relevant to this research inquiry. Getting to the historical root of today's responses found in the built environment is key to framing any potential responses to the future.

Theoretical Framework

Process tracing is rooted in the theoretical framework of scientific realism (Bennett & Checkel, 2014). Scientific realism maintains a splintered base of support but has, overall, been largely ignored (Maxwell, 2012). The theory that emerged in this dissertation, from the evidence reviewed, serves to advance understanding toward a knowable truth but does not assert it is the only possible construct of reality (Maxwell, 2012). Advancing toward understanding implies that scientific realism is based on a *gradual* progression toward scientific truths. There are 'different valid perspectives on reality' so any theory posited is likely unique to the researcher's perspective and in no way discounts other knowledge and observations that might emerge from analyses by others (Maxwell, 2012). Scientific realism suggests that the mechanisms and processes behind causality are real phenomena, whether observable or not (Maxwell, 1992). Scientific realism provides a useful position to explore this specific research inquiry because the data sources presented a rich variety of variables ranging from time, geography, culture, resources, intention, and individual differences. Also, the mechanisms identified by this analysis are unobservable, yet are manifested in the approaches used by communities of innovators and can be deduced through their actions and words.

Assumptions, Limitations, and Scope

Assumptions include that news reports in the popular press were rendered as an accurate reflection of reality as presented to the reporter, industry reports were reflective of genuine developments in the built environment, and that the scholarly and historical sources reflected true depictions of *communities of innovators*. In this assumption there is latitude that allows for the possibility that what was written can lean toward promotional content and positivity because innovation attracts advocates.

Limitations of this research include the researcher's personal bias developed from teaching environmental sustainability in business courses and developing collaborative, sustainability-oriented initiatives. Also, the Google Alert tool was used to capture daily news about innovation district progress, but on many occasions it was apparent this tool missed relevant developments. The alert tool was only specified to flag the specific terminology 'innovation district' although these types of developments are branded under a variety of phrases.

Another limitation is the omission of dozens of personal interviews with thought leaders that informed the research perspective but were conducted over a ten-year period without the advance approval of the Institutional Review Board at the university. In retrospect, had I realized so few academic publications would be available by 2015, these interviews over the previous ten years could have been documented to qualify as empirical research. Even to retroactively secure the approval process for that research, the knowledge gathered now would be irrelevant because those informal conversations were snapshots in time of a dynamic field that continues to evolve in real time. Also, those interviews occurred during multiple unstructured conversations over several days. In essence, the second iteration of those interviews would generate all new research and reduce the scope of the research to specific questions rather than broad inquiries experienced in the initial encounters.

The scope of the study restricted the research to a sampling of the thousands of *intentional communities* in the United States over the past five centuries, hundreds of research parks, and dozens of innovation districts. A few examples stretched the scope to nineteenth century England and twentieth century Italy, but for the most part, the analysis was derived from US sources. But more important than the sample size are the themes that emerged throughout the various forms of *communities of innovators*. Broad themes arise above the unique details and variables of thousands of communities, thus the appropriate view for theory development is best suited based on an overview perspective from 50,000 feet. Historical books sometimes covered individual communities but often they did so in a survey fashion so an overall impression could be conveyed rather than an individual profile. Because scholarly research on innovation districts is just now emerging, a Google Alert notification service was used from 2013-2016 to track real-time developments. Geographic focus was restricted primarily to United States examples.

Generalized findings of this study are useful to expand the understanding of those tasked with creating or fostering innovation. The mechanisms identified in these *communities of innovators* are pervasive and have already proved to be vital in many settings such as design work, patents, software development, and the development of Wikipedia (Leadbetter, 2008).

This broad application supports the premise that the findings can be generalized into an emerging theory and applied across a wide audience. This century is already being labeled as the era of innovation (Gassmann, Enkel, & Chesbrough, 2010). Speculatively speaking, this new emerging theory may be applicable to individual self-improvement, interpersonal relationships, organizational change, new product development, university positioning, and economic development strategies.

Definition of Terms

AASHE – *Association for the Advancement of Sustainability in Higher Education*

AIA – *American Institute of Architects*

APA – *American Planning Association*

AURP – *Association of University Research Parks*

EPCOT – *Experimental Prototype Community of Tomorrow*

GDP – *Gross Domestic Product*

LBC – *Living Building Challenge*

OECD – *The Organisation for Economic Co-operation and Development*

RTP – *Research Triangle Park*

TTO – *Technology Transfer Office*

ULI – *Urban Land Institute*

UN – *United Nations*

USGBC – *United States Green Building Council*

ZEH – ZEB – *Zero Energy Home and Zero Energy Building*

Summary

Sustainability is *the* key challenge facing humanity in the twenty-first century because of ecological instability, social equity unrest, and economic systems that operate outside the boundaries of sustainability principles, which require managing infinite resources and accounting for the impact associated with externalizing or delaying the true costs. Higher education institutions are in a unique position to respond to this imperative (Cortese, 2003). As of 2000, an estimated 6.7% of the global population held college degrees so the opportunity to influence is clearly available (Barro & Lee, 2000). Leaders in higher education must effectively manage their universities in the present in a rapidly-changing landscape, while simultaneously preparing for change in the future to be able to position and perpetuate their institutions (Flynn & Vredevoogd, 2010). If higher education institutions are to accept the challenge of inventing a sustainable society, then a paradigm shift in thinking will need to manifest in the thoughts, values, and actions of university leadership.

The research question driving the study is "What are the mechanisms historically used by *communities of innovators* as identified in *intentional communities* and in *places of innovation* that were used to approach their goals?" The inquiry was then tailored for the university setting. In the university context, the question driving the investigation is, "How can these mechanisms be applied to the environments created by higher education institutions so they can successfully fuel innovations that advance sustainability?"

A qualitative methodology was applied to conduct a historical analysis of past *communities of innovators.* Process tracing was the technique used to identify mechanisms and generate an emerging theory of how to foster *sustainability-oriented innovations* within the higher education system. Understanding how previous communities innovated their solutions can help university leaders formulate their response to future opportunities to achieve intentional outcomes. The future of humanity depends on our collective ability to become a global *community of innovators* where each member does whatever they can to propel today's civilization toward a sustainable condition. *Sustainability-oriented innovations* require not only invention, but also implementation and acceptance by consumers and institutions and that too falls in the domain of higher education (Rennings, 2000). Sustainability, few realize, is not optional and it certainly is not automatic.

CHAPTER II

REVIEW OF LITERATURE

This literature review contained over 1,200 academic articles, dozens of case studies, and hundreds of books—all which serve to elicit Albert Einstein's observations about projects becoming big and unwieldy: "Any intelligent fool can make things bigger, more complex, and more violent. It takes a touch of genius – and a lot of courage – to move in the opposite direction." As resources in the multiple literature reviews expanded, piles of articles containing data were generated so the challenge became to remain focused on distilling the essence, which gradually evolved into recognizing commonalities.

The investigation of the research questions focused on *communities of innovators* who had historically crafted creative and effective responses to their situations. Community is defined as "dense, multiplex, relatively autonomous networks of social relationships … and a mode of relating" (Calhoun, 1998). The *communities of innovators* investigated were geographically bounded to a place: a village, a region, a research park, an industrial area, etc. Their social response to new pressures and new possibilities is found in the history of *intentional communities*. Their economic response is found in the history of *places of innovation*. Both groups have sub-types that reflect ecological responses as well (J. Miller, 2015; Pickerill, 2015; Rennings, 2000). All of these *communities of innovators,* whether living in a community or working in industry, were driven by the idea of progress, which is considered key to perpetuating a society that strives for social improvement (Sargent, 1994).

Utopianism heavily influenced the experimental communities of Colonial America and all the way through U.S. history to the communities and developments seen manifesting today in the built environment. The investigation of utopian thinking, as it manifested in *intentional communities,* covers experimental communal settlements, communes, ecovillages, academic ecovillages, and innovation community prototypes. The investigation of innovative thinking, as it manifested in *places of innovation,* covers industrial districts, clusters, research parks, innovation districts, and universities. Utopianism requires innovative thinking and innovation is a manifestation of utopian beliefs that life could be improved through reimagined products and processes. The investigations of relevant literature and evidence were conducted under the belief that insights about past utopian efforts and innovative thinking could be used to inform the strategy of higher education's pursuit to foster *sustainability-oriented innovations.*

Intentional Communities: Organization of the Literature

The investigation of *intentional communities* produced evidence that is organized under two subheadings: utopian thinking and *intentional communities*. Under utopian thinking, the two areas covered are the contribution potential for higher education and the contribution potential for the built environment. Under *intentional communities*, the five areas covered are experimental communities, cohousing, ecovillages, academic ecovillages, and innovation community prototypes.

Utopian Thinking and Higher Education

With the literary contribution of Thomas More's book *Utopia* in 1516, the educated class was given a glimpse of an entirely new imaginary society, via satire, about fifteenth century science and religion (Kraftl, 2007). The word *utopia* was an invention by More as a pun on the Greek word *uetopia* meaning 'good place.' More removed the "e" but left the "u" meaning "no" and effectively created the term *utopia,* which translates as 'no good place' (Sargent, 1994). More's story of his utopian society was designed to provoke and unsettle; it introduced a tension between, on one hand, comfort and perfection, and on the other hand, the unsettling and unachievable (Kraftl, 2007).

After More, a genre was created when books were published that articulated utopian visions of a better, more egalitarian future (Sargent, 1994). They were published by recognizable names such as Sir Francis Bacon, Edward Bellamy, H.G. Wells, Aldous Huxley, and B.F. Skinner as well as hundreds of other authors. Even the earliest of utopia novels contained themes of social justice, environmental stewardship and economic growth (Harlow, Golub, & Allenby, 2011). In the early 1800s, there was a wave of utopian settlements built by a communal impulse to invent a utopian prototype in the United States (Schafer, 1978). So prevalent was utopian thought and utopian literature that it translated into planning actions and hundreds of intentional community settlements. Ralph Waldo Emerson commented in an 1840 letter to Thomas Carlyle that, "We are all a little wild here with numberless projects of social reform. Not a reading man but has a draft of a new community in his waistcoat pocket" (Carlyle & Emerson, 1884). Even then, designing a completely new society was an intellectual exercise of the common man and that exercise lingers today in the imaginations of futurists, planners, and inventors (Basiago, 1996). In the latter 1800s, dozens of literary utopian novels were published "like echoes of the actual social ventures of the preceding generation" as the dramas once lived became dramas penned as 'paper dreams' (Schafer, 1978). For example, Bellamy's utopian novel *Looking Backward,* published in 1888, sold over 400,000 copies in less than a decade (Claeys, 2011). In the twentieth century, utopian novels continued to be published and utopian communities continued to be planned and built but at a slower pace than the previous century (A. E. Bestor, Jr., 1953). As a warning signal against social ills that go unchecked or unresolved, dystopian novels began emerging in the early twentieth century as "narratives of decline" and in a noticeable trend, now displace and outpace the popularity of utopian film and literature (Vollrath, 2012a). In Vollrath's doctoral thesis on "the image of the future in the age of sustainability" he explains:

> The crises of modernity have intensified to such a degree that it is not even possible to fantasize about escaping them. Utopia is no longer amenable to the cultural imagination of the West, and it has not been since at least the end of World War II. (Vollrath, 2012b)

Fredrick Polak, one of the founding fathers of future studies, suggested that the concept of utopia was absolutely vital for society to perpetuate itself because images of the imagined future had the power to pull society toward that future as if utopian images functioned as a self-fulfilling prophecy (Polak, 1961). Polak worried that a world without utopian imagery would result in a void that would create a lack in humanity and end all progress made by Western civilization. He goes so far as to argue, "If Western man now stops thinking and dreaming the materials of new images of the future and attempts to shut himself up in the present, out of longing for security and for fear of the future, his civilization will come to an end. He has no choice but to dream or die, condemning the whole of Western society to die with him" (Polak, 1961). If true, then this makes eutopia (images of the future) "one of the most important artifacts devised by the human race" (Sargent, 1994). Sargent described utopia as a "distorting mirror in reverse showing how good we could look" and he suggested people are rightly disturbed by the concept of utopia because it suggests that both the life we lead and the society we have are "inadequate, incomplete, sick" (Sargent, 1994).

While utopian thinking seems fundamental to the spirit of innovation, the topic has remained relegated to isolated scholarly fields and has not been explored into the dialogue of those tasked with fostering innovation in either industrial settings or in educational venues. Utopian scholars describe utopian literature in great detail and context, but interdisciplinary scholars have had limited success in introducing that work to other fields for broader application (Goodwin, 2012; Stillman, 1990). The literature is almost devoid of how to adapt the mechanisms behind utopian thinking for application to the woes of today's civilization, though some have suggested utopianism as a "practical political philosophy" to "facilitate wise human actions" (Goodwin, 2012). If we accept that "sustainability is ultimately utopian in nature" then utopian studies offer useful untapped resources to sustainability advocates (J. P. Lockyer, 2007). Social ecology scholar Daniel Chodorkoff assures, "Utopian thinking today requires no apology. Rarely in history has it been so crucial to draw on the imagination in order to create radical new alternatives to virtually every aspect of daily life" (Chodorkoff, 1995). In consideration of higher education's interest in innovation "utopia caters to our ability to dream, to recognize that things are not quite what they should be, and to assert that improvement is possible" (Sargent, 1994). Given that one of the main roles of higher education institutions is preparing students for the future, introducing utopian thinking to scholars, students, and advocates of *sustainability-oriented innovation* could provide the necessary inspiration to give confidence to future generations of problem solvers.

Utopian Thinking and the Built Environment

Utopianism persisted in various forms throughout the development of the United States both in terms of literary contributions and physical examples of experimental communities built. It manifested as utopian socialism in the nineteenth century with a peak from 1820 to 1850. After the Civil War and World War I, discontent and the desire to start anew fueled more experimental communities. Spawned by the Great Depression, cooperative living experimental communities emerged. Utopianism was present in Koinonia Farm, a 1,000-acre farm near Americus, Georgia, which existed from 1941 to 1993; it began as a utopian demonstration of peaceful interracial integration and later served as an example of the legitimacy of the Civil Rights movement (Claeys, 2011). Koinonia Farm also practiced income and possession sharing; its members ranged from those who were illiterate to those possessing doctoral degrees (Winthrop, 1962). Incidentally, Koinonia Farm also spawned the home financing model for Habitat for Humanity. Utopianism was a driving force behind the communes of the 1960s and the ecovillage movements in the 1970s (Sargent, 1994). The United States was founded by utopian efforts that have remained omnipresent throughout its development and that continue today under new auspices.

In a sense, traces of utopianism can be seen in New Urbanism, an architecture and planning movement that started in the 1980s, specifically in its quest for increased quality of life, communal spaces, and an enriched public realm (Talen, 1999). More recently, sustainable urbanism, agricultural urbanism, and 'transition towns' have captured the imagination of planners designing for sustainability (Aiken, 2012; Farr, 2011; Qingji, 2002). A utopia for entrepreneurs can be seen broadly in the start-up community philosophy that dominates the *entrepreneurial ecosystem* of Boulder, Colorado (Feld, 2012). These modern examples are closely integrated into mainstream society, rather than standing separately as a radical form of social commentary of some desired alternative society as communal settlements sometimes do. These present day remnants of utopianism are positioned as incrementally better alternatives to conventional development but remain almost unrecognizable as utopian efforts because they do not use the language of utopianism or cite the principles of utopianism. It is generally agreed the 'idea of progress' inspired by utopian thinking 500 years ago by Thomas More has been on the wane the past 100 years (Stillman, 1990). But if these mainstream settlements are considered as loosely defined collaboratives of *communities of innovators*, then the similarities between today's modern communities and historic *intentional communities* are more obvious: they are place-based, micro-civilizations based on communally-agreed goals to create a more prosperous economy and a more livable society. Even while vestiges of our utopian heritage still remain, the field of utopian studies is not common in the conventional discourse of economic development strategists, architects, or planners of the modern built-environment designed to foster innovation.

Intentional Communities and Higher Education

To appreciate the relevance that *intentional communities* offer, it is necessary to examine the social origins of these communities and then establish definitions and labels. The origins of these communities can be traced back to the Diggers and Levelers, a group that led an insurrection on behalf of the common people during the English Revolution. In 1649, Gerrard Winstanley led a group of followers to a piece of common land at St. George's Hill and they claimed it for the common treasury so they could grow food cooperatively (Hardy, 2000). In 1652, Winstanley wrote the text *The Law of Freedom in a Platform* that would, centuries later, become instrumental and inspirational to socialist proponents Karl Marx and Frederick Engels (Hardy, 2000). In the short-term though, the Digger's communistic utopian effort lasted a matter of months and ended in failure, yet the ideology of the Diggers continued to resonate with the oppressed and disadvantaged. The American colonies were founded by Europeans influenced by these new ideas about how to organize and operate a democratic society and an equitable economy (A. E. Bestor, 1950).

The earliest settlements in pre-Colonial America can be defined as *intentional communities* because they were intentionally and purposefully designed to be alternatives to the existing society that the founders left behind in Europe. *Intentional communities* in the United States *officially* dates back to 1663 when a religious group from Holland, known as Mennonites, founded Swanendael in present day Delaware (Claeys, 2011). As a scholarly concept, *intentional communities* has a defined history of communal movements since Colonial America (A. E. Bestor, 1950). In general, *intentional communities* are divided into the secular and the religious (Sargent, 1994). Early secular settlements were entrepreneurial ventures such as Jamestown, established in 1607 in present-day Virginia; the intent was to fulfill the corporate charter by profiting from exports and eventually, the secondary goal became the sorting out of a more functional society. Other early settlements, known as *withdrawn communities* such as the Pilgrims at the Plymouth Colony of 1620, sought venues that provided freedom to practice their own religion. During the Reformation, Radical Protestants argued that the concept of private property was against the teachings of Christ. It was this philosophy of sharing equally that became entrenched in the ethos of generations of both religious and secular *intentional communities*.

An intentional community is a group of people who live together by choice to pursue an agreed upon goal (McLaughlin & Davidson, 1985). Dozens of labels exist to attempt to describe all the variations of *intentional communities*: communes, collectives, cooperatives, experimental communities, communitarian movements, *intentional communities*, intentional societies, practical utopia, utopian societies, utopian experiments, communal experiments, alternative societies, communistic societies, socialist colonies, elective communities, withdrawn communities, mutualistic communities, collective settlements, ecovillages, and *intentional communities* (Sargent, 2012). Fortunately, a simple definition of an intentional community suffices: "a group of people who have chosen to live (and sometimes work) together for some common purpose beyond that of traditional, personal or family ties" (Sargisson, 2002). An expanded definition includes the notion that the group must also have some kind of economic commonality: either they practice material sharing or participate financially in communal ownership of property and assets (T. Miller, 2010). There exists a complete discussion of modern intentional community definitions that includes sixteen perspectives from leading communitarian scholars in Miller's article "A Matter of Definition: Just What Is an Intentional Community?" (T. Miller, 2010).

As early commonwealth settlements in the colonies stabilized and grew over time, they lost their communal origins and became capitalistic towns with private property rights; however, a diverse utopian thought remained a persistent inspiration in the psyche of a young United States. The generation who inherited Colonial America's novel new republic, and its lofty goals set forth in the Declaration of Independence, continuously reimagined a more optimal society. One such effort was led by Englishman Robert Owen, known for introducing radical new social practices to his mill operations and for coining the term *socialism*. He purchased New Harmony, Indiana, a vacated frontier town, and built his experimental community by advertising for the leading thinkers and the communitarian-minded to occupy what was conceived as 'a village of philosophies' (Schafer, 1978). The first citizens were a very educated lot; they traveled to New Harmony by boat via the river and inspired the cliché 'a boatload of knowledge'. The intention was to establish the community with an inaugural group who were wise enough to execute the founding fathers' visions (Carmony & Elliott, 1980). Owen's goal was to "demonstrate a true community of diverse intellectual types could coexist and prosper" (Schafer, 1978). New versions of *intentional communities* continued to emerge as potential prototypes of idealized societies and their numbers peaked when 100,000 citizens participated by living in experimental communities during the westward expansion of the United States in the nineteenth century.

In contrast to dreamy utopian writings, the real experimental *intentional communities* were "prodigious feats of consistent social and physical design" distinguished by "imagination and inventiveness" (Hayden, 1976). The dwellings of the Shakers were steeped in "natural lore of earthly paradise, frontier self-reliance, democracy, and moral superiority" (Hayden, 1976). Such communitarians described their settlements as 'inventions,' referring to the analogy of a mechanical invention that could be designed and mass produced (Hayden, 1976). "These examples provide us with substantial experience of the rewards and problems of building for a more egalitarian society. Any group involved in environmental design, as part of a broader campaign for societal change has much to learn from them" (Hayden, 1976).

Fast forward to the twenty-first century and consider that higher education has been on a quest for interdisciplinary orientation for about half a century (Klein, 1990b). The earlier experiments in communalism were a quest for harmony, which meant that "a harmony of person with person, of humanity and nature – a fusion and total synthesis of nature and culture, human society and the basic organization of the universe" (Schafer, 1978). It would seem higher education institutions have much to gain by considering how *intentional communities* relate to the development, improvement, and responsiveness of their industry:

> The communalists formed a strong, steady force in American life, and we owe to them a great deal of our pluralism, the toleration for lifestyles and ideologies apart from our central bourgeois institutions. The communalists forced on nineteenth century America the example of functioning groups not based on laissez faire economics, secular utilitarianism or corporate greed. They were a constant reminder of social, religious and political ideals, a kind of conscience, however unwanted, for expanding, westward-loving America. Their existence was a safety valve for American politics—an alternative reality in which many republican virtues were kept alive, which reminded America of its multifaceted religious heritage, which pressed the issues of freedom and genuine social equality and a radical esthetics of life. Theirs was a counterbalance to the complacencies of the age, the other side of America to slavery, Indian wars, exploitation of immigrant labor, economic suppression of the yeoman farmer, etc. The force of utopia was one that kept alive much of the faith in America through imperialistic wars, the fight for the union and the rampant expansionism of industrial technology. We should not easily forget the dreamers and doers who maintained a steady American vision in a troubled age. (Schafer, 1978)

Institutions of higher education take responsibility for cultivating the next generation of citizens not only through conventional curriculum but, increasingly in the past few decades, through extending educational opportunities to include dormitories referred to as "living learning communities" (MacGregor & Smith, 2005). This is a step toward viewing student housing as a legitimate community, but what is missing from the academic discourse is how a university could expand their view from providing basic student housing to embracing their potential to provide a genuine community imbued with specific intentions much like what Schaefer refers to as 'dreamers and doers who maintained a steady American vision in a troubled age'. This echoes the earlier call for community put forth in the Harvard Graduate's Magazine in 1905 by President Lowell of Harvard University:

> We are come to the parting of the ways, where we must either make up our minds that the social life of the students is none of our affair – and in that case we had better probably better give up the college as an institution altogether, and confine ourselves to the work of the schools which prepare men for practical life; or we must bring our men [and women] together into a real community, with a common life – a true college life. (Ernst, 1904)

The potential purpose of what a 'real community' in the twenty-first century could be is the basis for the exploration of this study. The next five sections explore additional research gaps identified in five subsets of *intentional communities*: experimental communities, cohousing, ecovillages, academic ecovillages, and innovation community prototypes.

Experimental Communities

Whether historic or modern manifestations, "intentional communities have served as society's research and development centers for more than 250 years" (Kozeny, 2003). Celo Community was founded as an intentional community in the mountains of western North Carolina in 1937. It was based on communal ownership and land stewardship; its original intention was to be a master community "to be emulated far and wide" (Hicks, 2001). After Celo went through a deliberate ecological reorientation in the 1960s, it served as an inspiration to a wave of ecovillages in the 1970s. In anthropology, *intentional communities* are regarded as rich research because they "are instructive regarding the indigenous critique of the larger society" (S. L. Brown, 2002). It is only through self-awareness and self-critique that society can examine the areas failing to meet their potential, and take steps to address them.

Even communities labeled as 'failed experiments' continue to serve as a social beacon for alternative living. Stories abound about the far-reaching influence of residents who were part of the short-lived community of New Harmony, Indiana in 1825 (Denehie, 1923). Legislation and educational reform across Indiana, and the nation itself, was traced back to New Harmony residents. Two notable activists were Josiah Warren and Frances Wright. Warren established economic and social principles of anarchism a generation ahead of his time, and Wright, known as the United State's first feminist, founded a colony in Tennessee to emancipate and educate Negro slaves, decades before the abolitionist movement became established (Schafer, 1978). Even short stints of living in a social experiment can have deep impacts on residents and profound impacts on society's evolution. In this sense, New Harmony was a vital stepping-stone in the cultivation of new civic innovations that spread immediately; it was not a failure but a necessary iteration of an idea.

In the 1960s and 1970s over 700,000 people experienced life in communes in the United States (Schafer, 1978). Though most communes were short-lived and fraught with financial challenges and labor division issues, they were based on a communal and cooperative ethic. This practice of sharing later resurfaced as early software developers formulated the philosophical basis for *open source* software movement. Quite literally many of the programmers had once lived in 1960s communes and they carried forward those communal experiences with them to their future ventures (Leadbetter, 2008).

The value of the social experiment is also evident in the alumni of Wisconsin's Experimental College of the 1930s. The report of one student, sixty-five years later, captured the experience common to many about living in the midst of an innovative social experimentation in higher education. He described the program and living arrangements as "a central formative factor … besides the obvious aspects, but also the pervading emphasis on the moral and civic goals of the intellectual and social life of a democratic society" (Meiklejohn, 1932). A recent doctoral dissertation explored the changes experienced by alumni who had lived in experimental communities. It found that transformed identities and worldviews were the most common takeaways from living in a social experiment (Bochinski, 2016). These 'empowered utopians' report that upon leaving the experimental community, they have been able to use their new abilities to 'cultivate social connection' in the workplace (Bochinski, 2016).

Students who lived in the rare novel housing or social experiments on campus, and people who lived in non-academic experimental communities, show capacity for generating a lifetime of positive impacts on the larger civil society, yet such tremendous opportunities found in the past are rarely reflected in housing currently provided on university campuses today.

Cohousing

The rapid growth of cohousing in the United States over the past few decades signals a longing for a specific blend of communal lifestyles and private housing. Cohousing is a specialized type of intentional community that represents an evolution of conventional residential development. There are now in excess of 9,500 households that exist in cohousing neighborhoods in the United States (Williams, 2008). There are an estimated 123,000,000 households in the United States in total; cohousing remains a very small fraction of the existing housing stock. Most cohousing developments allow for private ownership of townhouses or single-family homes but mandate joint ownership of communal areas. The initial homeowners typically assume the role of developer and handle land acquisition, design, and financing. The purpose of cohousing is to provide a mutually supportive lifestyle to share the joys and divide the grief among neighbors who function as an extended family (McCamant & Durrett, 1994). Cohousing is also marked by an unusual neighborhood collaboration process beginning in the preplanning stages and carried forward to the daily governance. As a settlement type, cohousing stands as an innovation of form and function (Mark, 1991).

The universities of today have accepted that complex transdisciplinary problems are dealt with most effectively by collaborative teams (Lang et al., 2012; Nicolescu, 2005). The need to build collaborative capacity in the cohousing model mirrors the challenges universities face in building interdisciplinary collaborative capacity on campus. The potential research contribution is drawing those parallels so lessons learned in one form can be applied to the university community.

Ecovillages

Ecovillages are the blend of a social response and an ecological response; they come in a variety of forms. Some look like mainstream developments in suburban subdivisions or urban apartment complexes, while others reflect a rural, agrarian, or even primitive village approach reminiscent of the 1960s hippy communes. Ecovillages are a sustainability-oriented subset of *intentional communities* focused on green lifestyles and usually have a spiritual growth component (Sargent, 2012). As of 2010, Chitewere cited the Global Village Network as listing 347 ecovillages worldwide; ten of which were located in the United States (Chitewere, 2010).

By loose description, an ecovillage is "formed when groups of people choose to live with, or near enough, to each other to carry out a shared lifestyle with a common purpose" (Metcalf & Christian, 2003). The term ecovillage was first used in 1975 by the editors of Mother Earth News magazine to describe their experimental station for energy systems and organic garden experiments they built behind their corporate offices in Henderson, North Carolina. Ecovillages are built, in part, to provide a place to model sustainable lifestyle experiments.

Ecovillages are scientific experiments by virtue of the fact that a group of individuals collectively hypothesize an ecological model for society and then, using their own lives, test their ideas through real world experimentation (Sargent, 2012). Sustainability within an ecovillage is a transformative process that makes global sustainability more imaginable (Hong & Vicdan, 2015). Findhorn, the ecovillage in Scotland formed in 1962, is referred to as the 'frontier of sustainability' because it has a very small 'ecological footprint' (resource impact) and actively engages in educational outreach (Dawson, 2006). Other ecovillages also position themselves as demonstration models to inspire ecological lifestyles and consensus governance systems (Chitewere, 2006; Fischetti, 2008; Loezer, 2011). In 2007, the Department of Energy researched the energy efficient building systems used in construction of homes built in Wisdom Way Solar Village in Vermont (Aldrich, 2012). Wisdom Way was an experiment in energy efficiency within conventional building best practices, relative to the needs of conventional suburban neighborhoods, rather produced as an ecovillage development that serves as a holistic model of sustainability. Still, the DOE involvement does reflect a wider interest in building working models of some facets of sustainability.

Ethnographic research has allowed glimpses into the society of ecovillages and how they define sustainability (Castrejon Cardenas, 2007; de Oliveira Arend, 2013). The ecovillage research validates that a low-impact lifestyle is possible in an industrialized society and it also conveys the challenges and advantages of this form of communal living (Meadows, 1999). The book *Ecovillages: Lessons for Sustainable Communities,* written by a political science professor, offers tremendous scholarly contributions. It captures the lessons of fifteen ecovillages around the world as a collective source of inspiration for other sustainable communities. Although a university professor versed in sustainability wrote it, it did not explore how a university-owned community could incorporate ecovillage experiences into student housing to function as a model of sustainability.

While ecovillages seek to provide a balanced and holistic lifestyle, almost universally they struggle to provide for the economic foundation of the community (Walker, 2012a). Excessive consumerism is seen as the root cause for environmental destruction. Therefore, the concept of scalable capitalism is rather antithetical to the ecovillage culture (Baker, 2013). The concepts of bioregionalism, resilience, local economies, and eco-entrepreneurship have gained considerable traction over the past decade and these approaches may provide the venues to introduce business and economic principles into the ecovillage concept. A discourse in the field of Ecological Economics has broached the subject of using the urban ecovillage as a model for *degrowth,* even suggesting that *degrowth* itself might be a "concrete utopia" rendered in a "coherent picture" (Xue, 2014). Thus far, operating as a sustainable community in the greater context of an unsustainable society has proven problematic and divisive (Chitewere & Taylor, 2010).

Existing research that has attempted to address the operations of sustainable communities in a higher education context have not gone so far as to propose integration of the two models. For example, Kiernan Gladman's master's thesis in 2014 *Partnerships for Sustainability: Eco-Collaboration between Higher Education and Ecovillages* detailed specific examples of existing collaborations in Minnesota higher education institutions and ecovillages. It describes scenarios of potential collaborations but always on the basis of the entities remaining separate (Gladman, 2014). The idea of an ecovillage for faculty and staff on campus was explored briefly but only as a way to create a short-term immersion experience for students.

Whereas the colonial *intentional communities* were often withdrawn from mainstream society, ecovillages have from the beginning maintained an outward orientation toward greater society. The early ecovillages formed in the 1970s embraced concepts, technologies, and philosophies that were considered "quirky and irrelevant in a world of perceived energy abundance" but have increasingly attracted more attention from practitioners and policymakers in light of today's global environmental challenges (Dawson, 2013). There is even a gradual recognition that post-consumerist values and lifestyles will be forced to adjust to reduce resource use (Heinberg, 2010). Ecovillage research has shown that a high quality of life was present in *intentional communities* because social capital and natural capital were suitable substitutes for monetary capital so ecovillage residents were able to have a high quality life even with a lower than average income (Mulder, Costanza, & Erickson, 2006). Considering the economic paradigm for which students are educated comes through higher education institutions, the ecovillage model with its low-impact resource use and high quality of life supports university initiatives to educate an ecologically savvy populace.

Ecovillages that were once 'islands' are now networked with each other as well as with NGOs, government agencies, community groups, and universities (Dawson, 2013). They host visitors and internships and often partner with local education institutions. EcoVillage at Ithaca found collaborating with regional universities strengthened their community ties by furthering the mission of the EcoVillage and by augmenting curriculum for higher education (Allen-Gil, Walker, Thomas, Shevory, & Shapiro, 2005). In 2011, EcoVillage was awarded a $375,450 federal grant from the U.S. Environmental

Protection Agency (EPA) to fund "innovative on-the-ground approaches to creating dense neighborhoods that enhance residents' quality of life while using fewer resources" (Cosentini, 2011).

At present, universities, for the most part, produce innovation for economic growth rather than innovations that lead to sustainable equilibriums. Even though ecovillages are experimental and have many insights to offer sustainability-oriented innovators, a holistic research message about their potential to advance sustainability models does not readily exist and the available research stays in compartmentalized research fields where it has little chance of reaching the conscious awareness of university leadership.

The media's condemnation of Biosphere 2 in Arizona in the 1990s, a closed-loop ecosystem experiment, set a precedent that has discouraged large-scale, complex, ecological projects (Allen, Nelson, & Alling, 2003). In 1970, Arcosanti, another experimental settlement in the desert of Arizona, was built by architect Paolo Soleri and based on a term he created "arcology"- the fusion of architecture with ecology (Grierson, 2003). Soleri adamantly believed there was no way to predict the outcome of his experiment in social interaction so he referred to his project as an "urban laboratory" (Soleri, 1984). Soleri's settlement is a project of the Cosanti Foundation whose mission is "to build the urban laboratory Arcosanti envisioned so as to inspire research and foster cultural evolution that explores equitable and responsible relationships between cities and the earth's ecology" (Bochinski, 2016). Designed as a predecessor to the twenty-first century ecocity, it has yet meet its goal of housing 5,000 people but it is home to a small staff that hosts events and guests. More recently, Masdar City in Abu Dhabi was begun in 2008 as a zero-carbon green city, but the completion date of the $22 billion project has been extended to 2030 (Cugurullo, 2013). Masdar City has garnered much criticism as early models and prototypes often do (Lau, 2012). The idea of a university using one of their own residential communities as an experiment faces the challenging question of whether or not the stakeholders and residents would tolerate the planning complexities and the inevitable glitches of a prototype community based on cutting-edge, sustainability-oriented technologies or even simple ecological orientation. Research highlighting the *sustainability-oriented innovations* and the sustainability ethic as practiced in the ecovillage (a small scale to which a university could comfortably relate) could influence the willingness of higher education to experiment with providing model sustainable communities or ecological and social experiment venues.

Academic Ecovillages

The ecovillage concept first migrated into higher education in the 1990s. There have been exactly two universities that built ecovillage-inspired communities that blended residential student housing with sustainability research. The first successful effort to integrate sustainable living with academic research was the Lyle Center for Regenerative Studies at Cal-Poly Pomona, California. The project was spearheaded in 1994 by John T. Lyle of the landscape architecture faculty (Cal-Poly, 2015). Beginning in the 1970s, Lyle regularly challenged his students to design a community capable of living by only using resources on the site. By 1992, he and his colleagues had designed a curriculum and prototype community plan that successfully raised $4.3 million to build the Regenerative

Center. In 1994, twenty students moved into the housing. Upon the passing of Professor Lyle, the center was renamed in his honor. Though the center is not as comprehensive as originally planned in terms of building a permanent community, the research portion continues to be successfully integrated into the curriculum (K. Brown, 2015).

The second notable effort was driven by Larry Shinn, who was president of Berea College in Kentucky. As part of Berea's twenty-first Century Strategic Plan, it was determined additional housing was needed for the parents attending college and raising children. Shinn hired legendary green architect Sim Van der Ryn to design communal housing that incorporated sustainability research facilities and ecological design in the housing. The Ecovillage at Berea was built in 2007 and offers 50 townhouse-style apartments to the nontraditional students who are raising families. The cohousing-style townhomes highlighted ecological features such as passive solar design, rainwater catchment, and gardening plots near the front door. It was built for a cost of $10 million (Eilperin, 2005). A childcare facility is on-site to meet the needs of the sixty students in family housing. Additionally, an off-grid home is incorporated into the community and it is shared by four undergraduate students without children. The SENS House serves as a sustainability research venue and a demonstration facility for community outreach. The other sustainability research components showcased are a green building materials display, a permaculture garden, a working composting exhibit, and an aquaponics operation.

Another effort worth mentioning based on a sustainability orientation, but not labeled as an ecovillage, is Eden Hall, a rural satellite campus owned by Chatham University in Pennsylvania. Student dormitories were built on their new campus situated on a 388-acre operational farm; this campus also houses The Falk School of Sustainability. The 139-bed residential dormitories were completed in the fall of 2015, but it is not clearly articulated on the university website exactly how the housing interfaces with the sustainability curriculum of Chatham's Eden Hall campus and how it reflects ecovillage-inspired living.

A formal definition of an academic ecovillage has not been offered because the academic version differs significantly from a permanent community that is established and owned by private citizens. There was an attempt to describe the ecovillage form more specifically when Dawson listed five shared principles common to most ecovillages:

- They are not government-sponsored projects, but grassroots initiatives.
- Their resident's value and practice community living.
- Their residents are not overly dependent on government, corporate or other centralized sources for water, food, shelter, power and other basic necessities. Rather, they attempt to provide these resources themselves.
- Their residents have a strong sense of shared values, often characterized in spiritual terms.
- They often serve as research and demonstration sites, offering educational experiences for others. (Dawson, 2006)

Yet the idea that universities need ecovillages and ecovillages need universities drew out insightful observations from a PhD who specializes in study abroad courses that provide emersion experiences in an ecovillage (J. Lockyer & Veteto, 2013). Daniel Greenberg created a study-abroad program in 1999 through the University of Massachusetts-Amherst that took college students to ecovillages around the world (D. Greenberg,

52

2013a). Sue Gentile, the executive director after Greenberg resigned, wrote in the final edition of the Living Routes newsletter that between 1999 and 2014 they provided "1,485 students with the skills, knowledge, experience and wisdom needed to become social, cultural and environmental change leaders on local, national and global levels" (Redden, 2014). Greenberg articulated the symbiotic relationship they saw possible between the fusing of higher education and ecovillages in his article *Academia's Hidden Curriculum and Ecovillages as Campuses for Sustainability Education* (D. Greenberg, 2013a).

There are at least two other universities with notable advanced green building technologies incorporating into their residential housing. In 2011, Unity College in Maine built the first residence hall to feature Passive House standards; their Terrahaus reduced the heating load by 90% compared to conventional buildings constructed to minimum legal code (Fields, 2012). In 2009, University of California at Davis built West Village, the largest zero-energy community in the United States utilizing a 4-megawatt solar powered system. It is designed to house 3,000 people and offers 42,500 square feet of commercial space. West Village also headquarters uHub, a university incubator for innovations around sustainability (Kallushi, Harris, Miller, Johnston, & Ream, 2012).

These examples of academic ecovillages, or projects with components of advanced sustainable design, embody 'transformative sustainability learning' (TSL) by employing the organizing principle of the head-hands-heart, which integrates transdisciplinary study (head) with the practical skill of sharing and development (hands) with the translation of passion and values into behaviors or actions (heart) (Sipos, Battisti, & Grimm, 2008).

53

These examples of innovative communities likely have potential to provide transformative sustainability learning, but they have not been linked to the TSL theory because there has been no research published regarding the impact on learning outcomes of students living in an academic ecovillage. Each of these projects offers unique insights into their origins, planning, challenges, executions, and learning outcomes and all are rich areas for research.

Over the years, I continuously searched for published research and obscure unsung efforts. I informally queried scholars of ecovillage literature, utopian studies, and social innovations, about the idea of a university-sponsored ecovillage but was told the idea of a full-blown, academic ecovillage had been discussed in social circles for decades but never genuinely researched or fully attempted (D. Greenberg, 2013b; Longhurst, 2015; Sargent, 1994). There is a 2002 publication still available on a website about the efforts to network the advocates at various universities considering an ecovillage development. An undergraduate student named Yonatan Strauch from Mount Allison University in Sackville, New Brunswick Canada wrote a detailed synopsis of the various struggling efforts and suggested those groups network with each other to form an alliance (Strauch, 2002). Strauch's article mentions the ecovillage design class taught in 2003 by John Todd (inventor of the Living Machine), Professor Robert Costanza and others at the University of Vermont. More recently, it would seem the largest concerted effort to study how an academic ecovillage would fit within a university and design models occurred at Oklahoma State University over a 12-month period in 2012-2013 ("Wake Up & Dream project to host advocate of ecovillages across the world," 2012). Unfortunately, neither the experiences of the University of Vermont or Oklahoma State University were published in academic journals though the potential still exists for collaboration. All these efforts, especially the built examples at Berea and Cal-Poly, have potential to address the *knowing-doing* gap (Pfeifer & Sutton, 2000). There is a tremendous void in published research that documents these *knowing-doing* projects by universities and their attempts to explore combined models of sustainability with residential life, curriculum, outreach, research, and innovation.

Innovation Community Prototypes

Modern history offers several examples of attempts to establish technological and ecological advanced settlements. In Arizona, Arcosanti and Biosphere 2 were both built but have yet to reach their intended potential. Arcosanti was built and privately funded by its architect-founder Paolo Soleri and his foundation. Biosphere 2 was originally built by private investors, had a brief collaboration with Columbia University, and is now owned by the University of Arizona. Both entities are functionally operational. Since 1990, a planning philosophy called EcoCities has kept the conversation focused on sustainable city planning through annual conferences and global demonstration projects. Though discussion is beyond the scope of this dissertation, for a survey of sustainable cities theories and a sense of how constant this effort has been, review Basiasgo's 1996 article *The Search for the Sustainable City in 20th Century Urban Planning History* (Basiago, 1996).

Among the academic discourse scattered through various fields is a long forgotten call

from 1963 for 'scientific, intentional communities' as the venue to address the "slow

erosion of social idealism and the democratic social ethic' (Winthrop, 1963). Professor

Winthrop lived in the time of the Atomic Age and he recognized it would generate vast

amounts of science, chastised the "nature boys who are dreaming of pastoral idyll" and

those people "looking backwards to the peace, quiet, and relatively simplicity of the

social organic, agricultural communities of a prescientific age" (Winthrop, 1963). He felt

his concept for a micro community was 'a future eventuality' explaining:

> A micro community based upon new developments in science, technology, and
> invention – but which hopes to becomes an intentional community fusing the
> altruistic and social values of the religious impulse with a sense of individual
> dignity, worth, and participating in community processes inherent in the pristine
> ideas of face-to-face democracy – is, relatively speaking, a new idea.
> (Winthrop, 1963)

Absent from the academic discussion about sustainable communities, because they were

planned but never built, are two proposed projects that remain as examples of inspired

sustainable innovation thinking: Sim van Der Ryn's *Marin Solar Village* and Walt

Disney's *Project Florida* vision for a city of innovation.

The *Marin Solar Village* was designed to repurpose the decommissioned 1,200-arce

Hamilton United States Air Force base located 25 miles north of San Francisco. A

proposal to purchase the base and repurpose it featured a schematic site design by

Berkeley architect Sim Van der Ryn in 1979. *Marin Solar Village* was re-imagined as a

self-reliant, resilient community with the capacity to produce a majority of the resident's

food, produce 80% of its energy needs through solar energy collection, maintain its own closed-loop water recycling system, utilized electric cars, and provide on-site jobs for the 2,400 potential residents (Van der Ryn, 2005). The purchase proposal was defeated by local voters in a ballot election ("Marin County voters reject 'solar village'," 1979). The *Marin Solar Village* concept was a practical development designed to utilize the current technologies and innovations that were already within reach of any municipality seeking to redevelop a model community along ecological design principles. The original proposal did not include specific corporate or university involvement but did offer a viable living laboratory for ecological innovations on a broad community scale.

Walt Disney's vision for *Project Florida* included a new city oriented around community and innovation; it makes a very important, but largely unrecognized, contribution to city planning and to this dissertation by bridging the two concepts of *intentional communities* and *places of innovation*. Disney's vision has roots in Ebenzer Howard's prototype communities in the United Kingdom: Letchworth built in the 1903 and Welwyn built in 1922. Disney envisioned a state-of-the-art city that celebrated innovation, technology, and community. The concept model featured a radial/organic plan; a 50-acre town center enclosed by a mega-structure so the weather could be controlled; an internationally-themed shopping mall, a hotel and convention complex, office space, a greenbelt, high-density apartments, single-family residences; neighborhood centers, a satellite community, monorail, industrial district, and underground auto tunnels (S. Mannheim, 2012). In 1966, Disney unveiled his life's dream of building a new city where innovative American corporations would headquarter their R&D offices and invent life-enhancing technologies. He wanted this 'innovation industry' to be surrounded by residents who celebrated the spirit of American innovation and continuously implemented the various inventions into their daily lives. He stressed emphatically, this community was designed to cure the ills befalling American cities of the 1960s and, most importantly, bring people happiness (S. Mannheim, 2012). Disney called this new city in the heart of *Project Florida* the Experimental Prototype Community of Tomorrow. He stressed this was intended to be a prototype model to inspire other cities. He stressed that it would never be complete because it would be in a perpetual state of experimentation and invention that he referred to as state of '*becoming*'. A few months after unveiling his plan, Disney died. He behind models, drawings, plans, and speeches, but the visionary leadership necessary to execute this ambitious new evolution of civilization was absent.

In 1982, a technology and cultural showcase was eventually built in Florida by Walt Disney's brother, but that version of EPCOT was a shadow of Disney's original grand vision for a real city; it was a family-oriented amusement park to visit, not a place to live and not an industrial city center to foster innovation. Reminding society to imagine the future is vital for a vision to become a reality (Polak, 1961). Walt Disney wanted the people of his community to imagine the future and then experiment with the application of it while perpetually imagining the next future (Nachman, 2014). Transition management experts suggest envisioning sustainable future trajectories is vital in order to steer large-scale system innovations toward sustainability (Sondeijker, Geurts, Rotmans, & Tukker, 2006).

To fully appreciate the contribution potential of Disney's *Project Florida* requires an introduction to a Garden City, a prototype model community that was actually constructed in two locations in the United Kingdom at the turn of the twentieth century. Due to pollution from the Industrial Revolution, inventing a better community design for human habitat became a public health necessity in the late 1800s. Along with public health issues, urban squalor and social ills motivated Ebenezer Howard in 1889 to write a book that used design to propose a remedy: *The Garden City of To-Morrow: A Peaceful Path to Real Reform* (E. Howard, 1902). A Garden City was an early attempt to manage the increasingly complex industrialized city and create a hybrid innovation of town and country. A Garden City was proposed as the "best of town and the best of country" combined into the optimal blend and driven by visions of utopia (E. Howard, 1902). Ebenezer's proposal was in response to a time of discontent with industrialized cities. Mass production was seen as "the enemy of art, divorcing individuals from craft-based production and creating a new and inferior aesthetic" (Hardy, 2000). Howard's premise provided a restoration of a quality of life and meaningful existence seen as rapidly deteriorating.

A Garden City was more than just a planning response to environmental pollution; its underlying intent was based in social reform. Like many utopians before him and since, Howard believed a strategically designed urban form could foster cooperation *and* spiritual transformation. A Garden City was designed as a city of gardens, literally green space and greenbelts, for a self-contained community of 30,000 people. A Garden City also presented a new approach to land ownership that resulted in routing profits from land sales back to the commonwealth for public investment. Integrating a prosperous society with housing, industry, trade, education, parks, recreation, and transportation to create a resilient and equitable economy was—and still is—a monumental city planning task. Still, Howard was able to get two experimental towns built in the United Kingdom, both were completed in his lifetime: Letchworth in 1903 and Welwyn in 1920.

A careful read through Howard's text can detect the seeds of concepts that would later be recognized in the next century as fundamental sustainability precepts. Fair trade, social equity, and maximizing public good over private capital wealth were just a few of the topics Howard broached, though he lacked the terminology to name them. Howard attempted to solve the ills of the industrialized society with a holistic community design based in social reform and equity. Howard's design premise was driven by an underlying belief in balancing environmental resources, social equity, and economic stability; we recognize this today as the Triple Bottom Line of sustainable business (Elkington, 1997). The Garden City planning philosophy became a major movement in town planning in the twentieth century and a global phenomenon with many adaptations and interpretations, none of which remotely reflected the original intent of building wealth for the commonwealth (Hardy, 2005).

In the United States in the late 1920s, the Garden City model debuted as a new approach to town and neighborhood development. Several new neighborhoods were influenced by the Garden City philosophy with the most comprehensive effort begun in Radburn, New

Jersey in 1928. Radburn was ultimately became an unfulfilled promise that failed to be completely built as planned (Birch, 1980).

Although Radburn did not significantly impact United States development patterns—due to its timing around the Great Depression, not due to its relevance—the *Radburn Principles* of traffic segregation remain as a demonstrative example of a pedestrian-centric design that balances open space with the needs of residential housing and commerce. Despite many attempts to overlay the Garden City concept to thousands of new developments globally, only Letchworth remains as the "most complete example of an attempt to create a Garden City, not only in the United Kingdom but internationally, and certainly merits a study of this sort" (Hardy, 2005).

While some planning and architecture schools may expose their students to the deep significance of Garden City contributions, what actually gets built continues to fall woefully short of the ideal (Sharifi, 2016). This is because practitioners are hired by developers who are usually versed in finance more than social reform or sustainable development. As expressed in a preface written by F.J. Osborn for the 1946 reprint of *Garden City of To-Morrow*, "no book of significance has enjoyed less academic notice or prestige" (S. E. Howard, Mumford, & Osborn, 1946). Osborn laments further how "so few trained thinkers detect that Howard possessed extraordinary intuition and judgment" even though Howard was widely considered an inventor of solutions to complex social and economic problems (S. E. Howard et al., 1946). Even modern planning historians acknowledge additional prototype communities are not fully attempted because the "apparent simplicity belies the hidden complexity" (Hardy, 2005). As a new planning paradigm based on sustainable development, Garden City has been demonstrated to be well suited to the demands of the twenty-first century (Ward, 2005). Howard's Garden City may have looked like just another utopian panacea in the long tradition of experimental communities, but is, in fact, aimed at "the middle course between utopianism and pragmatism" (Hardy, 2005).

Howard's Garden City was a tremendous innovation in itself, but it also contains significant contributions when presented in the context of *places of innovation* because of its direct impact on Disney's design approach to *Project Florida*. Howard's ideas can find broader application in the design of environments that foster innovation, as can Disney's and Van der Ryn's. A university campus is often a town within a town. They consider their campus, employees, and residents to be a community very similar to a town. As such, universities have the potential to draw ideas from prototype projects that inspire them to reimagine how their university could serve as a model community that advances ecological technologies and *sustainability-oriented innovations*. Research has not yet compiled the historical examples of holistic developments dedicated to innovation within the context of a sustainable community that university leadership can use to craft a strategic mission involving modeling sustainability.

Places of Innovation: Organization of the Literature

The investigation of innovative thinking as it manifested in *places of innovation* covers industrial districts, clusters, research parks, innovation districts, and universities. Research about these *places of innovation* originates in several academic disciplines including economic geography, social sciences, urban planning, organizational change, business, innovation, and entrepreneurship to name a few. Although innovation is a popular research topic, there has been very little focus on the initial appearance of innovation itself (Ruef, 2002). Rather, research concentrates on the diffusion, adoption, and viability of given innovations (Ruef, 2002). Even the leading cluster researchers agree there is "relatively little work" outside of a few United States studies that "credibly identifies the role of local innovation in local employment growth" (Chatterji, Glaeser, & Kerr, 2013). The support for innovation is based on the belief it will lead to economic growth and thus economic stability, but economic growth without consideration for consequences can lead to unsustainable systems, a concern that is reserved for discussion in chapter 5. There is, however, an abundant amount of publications on clusters and industrial districts: 3,955 academic articles were published between 1957 and 2014 and revealed six sub-fields emerging in the literature (Hervas-Oliver, Gonzalez, Caja, & Sempere-Ripoll, 2015).

The research goal of this dissertation is to take a macro view of *communities of innovators* to determine what mechanisms that fostered innovation were reflected in the evidence reviewed. The organization of the evidence from these *communities of innovators* is arranged under three subheadings: 1) industrial districts and clusters; 2) research parks and innovation districts; and 3) universities. This evidence has a different quality than that found in the intentional community investigation. Some of it relies on the popular press and promotional material because the newer forms of *places of innovation* are still emerging research topics. Also, the academic literature is based on a broader perspective of conglomerated examples rather than individual case studies.

These *places of innovation* in the private sector are known by a plethora of names: cluster, industrial cluster, innovation cluster, knowledge cluster, regional innovation cluster, national innovation cluster, industrial district, I-district, industrial zone, new industrial district, innovation district, innovation network, enterprise zone, innovation zone, global innovation network, technopole, and territorial innovation model. Those in the university realm are referred to under the following names: innovation university, entrepreneurial university, innovation campuses, and experimental colleges. Associated with these dozens of nouns is a collection of multiple definitions for each noun (Hamdouch, 2007). One concept upon which the scholars do agree is that the disciplinary segregation has resulted in an array of definitions that vary from discipline to discipline, thus creating considerable confusion as to which definition to use with which term (Hamdouch, 2008). For this very reason, the research for this dissertation bypassed the semantics to look for common mechanisms in the approaches used in all these various venues to foster potential for innovation. Broadly speaking, the research gap that exists for higher education institutions is an unrecognized confluence of mechanisms used to foster 'innovation in place' ranging from the private sector industrial districts and clusters to the university partnerships in research parks, innovation campuses, innovation districts, and universities.

Industrial Districts and Clusters

To understand the innovation potential found in the ecosystems of places like Silicon Valley and Boston's Route 128, a discussion of nineteenth century industrial districts and twentieth century clusters is warranted. It is also useful to have a cursory understanding of industrial districts and clusters because they are predecessors of, and exist in parallel to, the other innovation venues in which universities participate.

Michael Porter of Harvard University did not invent, but pioneered, the concept of clusters (Porter, 1998). Porter defines clusters as "a geographical proximate group of interconnected companies and associated institutions in a particular field, linked by commonalities and externalities" (Porter, 2008b). Clusters were not a new idea but, according to Porter, are rather an extension and application of economist Alfred Marshalls' work in 1920, though not all scholars agree on this genealogy (Sforzi, 2015). Porter suggested that medium-sized clusters of firms were more competitive than large isolated manufacturing firms thus creating the premise behind the nineteenth century industrial district. The fact that Porter's modern cluster definition originated from Alfred Marshall's turn-of-the-century industrial district concept, and retains key elements of that concept, continues to create opportunity for terminology confusion (Hamdouch, 2008).

Porter's advocacy of the cluster concept, *his* cluster concept, has been likened to a marketing strategy to brand a product (R. Martin & Sunley, 2003). Martin and Sunley express concern that Porter's cluster concept is a construct that has taken license with 'cavalier' and conveniently 'loose' definitions; their paper cites ten additional cluster definitions from other authors (R. Martin & Sunley, 2003). Moreover, Sforzi, an Italian scholar versed in the original innovation district scholarship and theoretical frameworks, refutes that Porter's cluster concept and the industrial district share similar theoretical roots (Sforzi, 2015).

Industrial districts originated in the traditional artisan economies and manufacturing sectors in Italy in the nineteenth and twentieth centuries (Brusco, 1990). Examples include knitwear in Modena and Capri; clothes and ceramic tiles in Modena and Reggio; wool textiles in Prato; cycles, motorcycles and shoes in Bologna; buttons in Piacenza; tomato canning and ham production in Parma; leather tanning in Santa Croce; pig breeding in Reggio Emiha (Rogerson, 1993). To appreciate the complexity of the ecosystem that develops in a community around a product or industry, consider the wool textile industry in Prato. It has splintered into many specialized operations involving spinning, dyeing and weaving, which are also associated with dozens of additional steps to produce an array of finished goods. Several hundred specialized brokers and dealers contract the raw materials to subcontractors to create finished goods to market (Rogerson, 1993).

The industrial district is referred to as a "*community* of people who live and work in the same locality" (Boix & Trullén, 2010). Industrial districts are "geographically defined productive systems, characterized by a large number of firms that are involved at various stages, and in various ways, in the production of a homogeneous product" (Bagnasco, 1977). The term 'industrial district' was coined in 1979 by Giacomo Becattini and is now referred to simply as "the district effect" or "I-District" by scholars (Sforzi, 2015).

Although the industrial district originated based on local economic successes in Italy, the conceptual model is used internationally to understand how to structure policies that support industrial development planning (Rogerson, 1993). These best practices from the "Third Italy" constitute policy lessons useful in regional and small scale economic development (Boix, Sforzi, & Hernández, 2015).

In the first half of the twentieth century, these clusters or districts grew organically in response to communication and transportation technologies (Wessner, 2012). The industrial district concept is still recognized today and, in fact, has spawned an ongoing field of academic research (Boix et al., 2015; Sforzi, 2015). The research line also produced a new typology among United States scholars that recognized the New Industrial District (NID), along with four different kinds of districts: Marshallian, Hub and Spoke, Satellite Platform, and State-Anchored (Markusen, 1996). The Italian scholars dispute Markusen's application of industrial district framework in her analysis (Sforzi, 2015). The NID remains a developing concept in the quest to understand the district effect in regards to globalization (Xiaojian, 2011). A significant feature of industrial districts is that a very high proportion of these firms are small or very small (Pyke, Becattini, & Sengenberger, 1990). Scholars describe their findings as showing small-firm clusters are more innovative than even large firms with economies of scale advantages (Boix & Galletto, 2009). For decades it assumed as that industrial districts and innovation were correlated, but in 2006 an empirical analysis on patent generation was conducted on Italian firms within an industrial district and compared to those outside a district. Using patent generation as a proxy for innovation activity, the analysis confirmed that the correlation between innovation and firms located within an industrial district was stronger compared to innovation within those firms located outside the industrial district (A. Muscio, 2006). Further, it has been shown that university *proximity* to current industrial districts facilitates the knowledge transfer activities of a university and leads to increased private funding through industry collaborations (A. Muscio, Quaglione, D., & Scarpinato, M. , 2012).

Innovation cluster.

The terms industrial district, industrial cluster, innovation network, and innovation district have become increasingly interchangeable over the last few years even though there is an implied difference of scale (Clark, Huang, & Walsh, 2010). Modifying the term from industrial cluster to innovation cluster complicates the definitions further, yet innovation cluster has been able to elbow space for its own unique definition in this field crowded with duplicate terminology.

Due to disciplinary segregation, there are multiple disputes about how to analyze what factors underlie the emergence of innovation clusters (also known as innovation networks), how to structure those factors and foster organic evolution, and how to define these places by spatial boundaries or social networks. Abdelillah Hamdouch, a researcher in Paris, attempted to bring order to the chaos in the field by reducing the definitions as originating from just two schools of thought: one being led by Michael Porter in his 1990 book *The Competitive Advantage of Nations* and the other originating from the Organisation for Economic Co-operation and Development (OECD). In seminal contributions, Hamdouch provided a critical analysis of ambiguities and overlaps between definitions of innovation clusters that were generated by leading scholars (Hamdouch, 2008). He offered his synthesized definition of innovation cluster:

> An innovation cluster comprises an ensemble of various organizations and institutions (a) that are defined by respective geographic localizations occurring at variable spatial sales and within specific institutional environments, (b) that interact formally and/or informally through inter-organizational and/or interpersonal regular or more occasional relationships and networks, (c) and that contribute collectively to the achievement of all kind of innovations within a given industry or domain of activity, i.e. within a domain defined by specific fields of knowledge, competences, and technologies. (Hamdouch, 2008)

According to Hamdouch, a key feature of a cluster is that it is an alternative type of *vertically-integrated* value chain usually for a *specific* industry. Innovation clusters are self-organized and spawn innovation in their specific industry due to a critical mass of competencies that understand the technologies required in a given industry (Hamdouch, 2008). Innovation clusters exhibit strong inter-organizational complementarities from a diversity of actors embedded in an entrepreneurial culture (Hamdouch, 2008). Clusters are generally larger in geographic terms than an industrial district; clusters may encompass a city, a region, or even a multi-state area. Examples of clusters include the United States include: Colorado cleantech, Michigan battery cluster, and Wichita aviation cluster. The *U.S. Cluster Mapping Project*, sponsored by the U.S. Economic Development Administration and led by Michael Porter, identified 51 distinct clusters in the United States and these geographic identification designations are used as policy tools (M. E. Porter, 2014).

Prior to their advocacy of the innovation district concept, The Brookings Institute published a think piece in 2010 in support of innovation clusters; they stated clusters were the key to economic recovery (Muro & Katz, 2010). Shortly before Brookings's *New Cluster Moment* paper was released, a new cluster term was coined – regional innovation cluster (RIC) – by a white paper published by a think tank (Yu & Jackson, 2011). By 2011, the RIC concept was embedded into policy framework identified by the Obama administration as a venue to diffuse investment aimed at spurring economic growth through innovation (Yu & Jackson, 2011). Scholars responded by publishing new interpretations of clusters in the academic literature and thus legitimized the study of regional innovation clusters (Enright, 2003; Laperche, Sommers, & Uzunidis, 2010).

Research Parks

Places of innovation known as research parks, technology parks, or science parks are designed to act as conduits of knowledge transfer flows between the firms in the park and the universities for the ultimate purpose of economic development for the region (Link & Scott, 2011). Specifically, a *university* research park is defined as "a cluster of technology-based organizations that locate on or near a university campus in order to benefit from the university's knowledge base and ongoing research" (Link & Scott, 2005). The transfer flow operates two ways: the university not only transfers knowledge to private industry but "expects to develop knowledge more effectively given the association with the tenants in the research park" (Link & Scott, 2006).

The first research park was Menlo Park built in California in 1948. The most famous parks are the Stanford Industrial Park built outside of San Francisco in 1951 (subsequently known as Silicon Valley) and the Research Park Triangle in North Carolina initiated in1959. These early parks were sequestered away from the mainstream and built on the isolated suburban model. *The Manhattan Project*, a secret United States project to develop the atomic bomb in WW2, was considered effective in producing remarkable breakthroughs in science *specifically* because of its secret and sequestered location in Las Alamos, New Mexico. The strategic placement of this research facility in an isolated region had tremendous influence on the design of the early research parks.

The growth of research parks in the 1980s was greatly facilitated by the introduction of new laws and policy incentives (Link & Scott, 2006). The Bayh-Dole Act of 1980 introduced a new federal patent policy that provided financial incentives to bring to market innovations funded by federal grants. This meant it became legal for a university and the contributing researchers to hold title to an invention developed with public funds. The Research and Experimental Tax Credit of 1981 originally provided a 25% tax incentive to firms that increased R&D expenditures over those made in previous years.

The National Cooperative Research Act of 1984 encouraged the formation of research joint ventures among United States firms and universities thus allowing them to act as partners in those ventures (Link & Scott, 2006).

The basic premise of the public-private partnership used in the university research park model was articulated in 1988 when the *Triple Helix Model* was proposed by Professor Leydesdorff at the University of Amsterdam (Leydesdorff, 1988). When Leydesdorff collaborated on publications with Stanford Professor Etzkowitz, the *Triple Helix Model* of collaboration of university-industry-government relations was widely popularized as the appropriate model to facilitate technology transfer. The *Triple Helix Model* proposes that the university is uniquely positioned to provide the key to a dynamic system of innovation because it provides the continuous flow of curious students that are the 'life blood' to the innovation system (Etzkowitz, 2014).

After a decade of a global building frenzy of research parks in the 1980s, academics in the United Kingdom began to question the assumption that academic research and industry had strong linkages to economic development, and found there was little empirical evidence to support the relationship (Quintas, Wield, & Massey, 1992). The literature does not make a consistent case that the research park model is actually the most beneficial venue for achieving economic prosperity through university-derived inventions, but the sunk costs had already been incurred. Leading United States scholars have admitted economic research on science parks and university research parks has been lacking because the research field was in the 'embryonic stage' (Link & Scott, 2007). They posit that, because these public-private partnerships have multiple stakeholders with

72

varying goals that affect performance, growth models of the economic benefits are difficult to establish empirically (Link & Scott, 2007). The 2009 report from the Association of University Technology Managers (AUTM) disclosed only $2.3 billion of licensing revenue was generated by $53.3 billion in federally-sponsored research among 181 universities (Kanter, 2012). Given that industry growth is not based on empirical evidence, it possible the university research parks were built as a trend in response to the general global productivity declines and because the political culture that provided funding did not require justification. Innovation scholars in the U.K. concede that "the empirical evidence base for science parks' effectiveness as a policy intervention is sparse, mixed, and contradictory, though much of the research tends to suggest some degree of positive association" (Price & Delbridge, 2015). Currently, the lines have blurred between innovation districts, innovation campuses, and university research parks, so generating an accurate number is difficult. In an economic development handbook produced by a University of Michigan class in 2005, the authors quoted AURP's website as listing 195 research parks in the United States in 2005 (Ahn, 2005).

Research parks, like commercial strip shopping centers, have a predictable life span. They devolve unless effort is put forth to evolve them and extend their usefulness. *The Future of Knowledge Systems* publication identified 14 trends that transition experts used to create three scenarios and forecasted strategic implications. This research was specifically directed at the current group of 40 research parks in the United States that were over 25 years old in 2009 (Townsend, Soojung-Kim Pang, & Weddle, 2009). As an example of evolution, consider how the *knowledge economy* manifested in North

Carolina. In the 1950s, a coalition of leaders in North Carolina pioneered a bold concept in a calculated effort to retain their college graduates in their home state through an economic development strategy. It called for organizing a place of innovation where research labs could co-locate in a large rural area between three of the state's universities: Duke, the University of North Carolina, and North Carolina State University. This was the launch of the Research Triangle Park known now as RTP. It was a place where corporate headquarters and their innovators were invited to strategically position their operations close to three higher education institutions, so regional ties could evolve between industry and higher education. The RTP now boasts over 200 companies and employs 40,000+ people with technical expertise.

But even a very successful, high-profile park has a product life cycle and, by 2012, the RTP experienced vacancies that pressed them into exploring how to compete for innovative companies. One strategy identified was for RTP to redevelop holistically by providing a residential community with urban amenities such as coffee shops, retail, and entertainment. The Brookings Institute refers to the RTP as the 'urbanized science park' model of innovation districts (B. a. W. Katz, Julie, 2014). The specific amenities and features that a research park should incorporate to be considered an innovation district has not been determined because the innovation district form is still so new and not even industry standards have evolved. Each innovation district is developed individually based on the approach of the local stakeholders desires and the practitioners' expertise.

Innovation Districts

    Origins.

For the innovation districts involved with university commercialization, their history begins early in the twentieth century. Technology transfer offices (TTOs) started in 1923 when Harry Steenbock, a professor at the University of Wisconsin, invented irradiation of vitamin D, but the university lacked the infrastructure to patent and commercialize inventions so he created the Wisconsin Alumni Research Foundation (Litan, Mitchell, & Reedy, 2007). Now TTOs are changing their business models, driven in large part by the life sciences, and some are becoming engaged with *places of innovation* ranging from research parks and innovation campuses to innovation districts. The University of Pennsylvania is creating their new research park, Pennovation Center, in the 3,700-acre Lower Schuylkill Innovation District, which is projected to have a $63 billion economic impact on the area (Huggett, 2014). Wake Forest developed their Innovation Quarter on 145 acres and it houses 50 technology companies including 26 academic units for a total employment of 3,100 people; the development is valued at $600 million (Huggett, 2014).

The interest and activity is more pervasive and widespread than anyone in the industry realized or at least as disclosed openly in the popular press or the literature. *The Harvard Gazette* article in July of 2014 published an article on innovation districts that cited "more than a dozen United States cities have designated sections of their downtowns as micro business empowerment zones targeting the innovation economy" (Pazzanese, 2014). Another 2014 article published about the changing venues for technology transfer in higher education stated there were over 80 cities either executing plans for an innovation district or in preliminary explorations (Huggett, 2014). In less than half a decade, the innovation district development in the United States has grown exponentially from one in St. Louis in 2002 and another one in 2010 in Boston to over 80 in 2014.

The shift in urban governance from managerialism in the 1960s to entrepreneurialism in the 1980s was noted by human geographer David Harvey as a new paradigm worthy of examination and scrutiny (Harvey, 1989). The innovation district phenomenon is, in part, a manifestation of the new role of a city as seen by the entrepreneurial mayor. These city leaders see the city as part of the *innovation ecosystem* and/or the *entrepreneurial ecosystem*. An innovation district can be thought of as a confluence serving the mutual interests of a civic leaders seeking local economic development, a university seeking increased relevance through industry associations, and corporations seeking access to knowledge systems. While innovation district developments are collaborative efforts, an anecdotal review of the popular press reports indicates the initial impetus for leadership seems to originate from either the municipality or the university.

An innovation district is typically a *place of innovation* on the urban neighborhood scale and it is intentionally designed to foster innovation and entrepreneurship (B. a. W. Katz, Julie, 2014). The Brookings Institute was the first to publish a non-academic book about innovation districts: *The Metropolitan Revolution: How Cities and Metros Are Fixing Our Broken Politics and Fragile Economy.* The authors define an innovation district as "geographic areas where leading-edge anchor institutions and companies cluster and connect with start-ups, business incubators, and accelerators" in places that offer housing, office space, and retail that is all connected by sidewalks and transit (B. Katz & Bradley, 2013). They identify three types of innovation districts where metro regeneration is likely: anchor institutions (universities or medical schools), re-imagined derelict areas (downtown Detroit), and urbanized science parks (research parks with housing and amenities added). The broader goal of the innovation districts is to create a place "to spur productive, inclusive, and sustainable economic development" by addressing "growth, natural austerity and its local challenges, social inequity, sprawl, and environmental degradation" (B. a. W. Katz, Julie, 2014). The authors provide examples of innovation districts starting in the United States such as Cortex in St. Louis, Seaport Innovation District in Boston, and Research Triangle Park in North Carolina.

The scholarship.

While popularly accepted and implemented as a resource in the practitioner realm and policy circles, the works contributed by The Brookings Institute have been politely criticized by academicians for failing to use rigorous, empirically-derived data to support their premises (Pazzanese, 2014). Harvard economist Edward Glaeser observed that, "innovation districts are … a hypothesis; they're not a proven strategy at this point in time. I think they're as sensible a hypothesis as any one out there, but they're merely a hypothesis" (Pazzanese, 2014).

The scholarship around innovation district planning, development, and execution is truly sparse. There are multiple stakeholders in every innovation district, the most common ones being: the municipality, educational partners, the corporations, the entrepreneurs,

NGOs, and the community. To capture the true essence of an innovation district presents holistic challenges that are complicated by virtue of the fact the innovation district form itself is evolving and adapting to its unique environments. Research that puts the innovation district developments in context to the other public-private partnerships that universities engage with for commercialization is lacking. Research contributions that link the sustainability imperative to innovation venues are also sparse but the field of *sustainability-oriented innovation* (SOI) does have a presence across many fields. In a recent review of 100 articles published between 1992 and 2012, researchers found the publications were spread over 55 different journals (Adams, Jeanrenaud, Bessant, Denyer, & Overy, 2015).

Innovation districts, as a distinctly branded economic development tool, are less than a decade old in the United States and, as such, the scholarship around them is still evolving. The people intimately familiar with the development process are the Mayors' offices and the practitioners in architecture and planning firms. To date, there has been scant academic attention paid to the practitioners who plan innovation districts. Students in an urban planning course taught at the University of Texas produced *The Austin Anchors & The Innovation Zone: Building Collaborative Capacity*, a 129-page report examining the collaborative process of the early planning phases (S. R. Greenberg, 2015). The report includes the following case studies: Central Keystone Innovation Zone in Pittsburgh, Cleveland Health Tech Corridor in Cleveland, CORTEX District in St. Louis, Fulton Market in Chicago, Kendall Square in Cambridge, Mission Bay in San Francisco, South Boston Waterfront, South Lake Union in Seattle, Texas Medical District in Houston, and

University City in Philadelphia. Aside from the Brookings book, the only other book available on innovation districts is *Innovation Districts: A Toolkit for Urban Leaders* (Morisson, 2015). This same author who published a master's thesis that developed a framework to compare innovation district developments in Barcelona and Boston (Morisson, 2014). In 2014, there was also a master's thesis published that made policy recommendations about how to expand the initial innovation district in Boston to surrounding neighborhoods (Taylor, 2014). The American Institute of Architects (AIA) published *Cities as Labs*, a glossy 74-page compilation of innovation district profiles, innovative housing, resiliency resources, and public space projects as an industry case study in support of innovation venues in the built environment (Rainwater, 2013). Rainwater refers to innovation districts as places using 'hyper-placemaking' strategies that build new relationship infrastructures based on linking assets in *proximity* (Rainwater, 2014). The only critical assessment about innovation districts has come through a sprinkling of popular press articles that are more skepticism than scholarship (Russell, 2014; Winkler, 2014). Though they offer voluminous anecdotal objections, the opinions can serve to inform academic investigations and stir thoughtful debate among practitioners and policymakers.

A new variation of innovation district, EcoInnovation districts, evolved by layering goals addressing environmental, economic, and social equity. EcoInnovation districts are in planning phases in Boston, Massachusetts as well as Pittsburgh and West Oakland in

Pennsylvania ("EcoInnovation district initiative," 2014; "EcoInnovation District Uptown Oakland," 2016; J. Miller, 2015). The EcoDistrict framework was piloted in these cities in 2015.

The innovation district model explores the *physicality* of how innovation networks are overlaid in a specific geographic location (B. a. W. Katz, Julie, 2014). Research on "third places" focuses on spatial qualities of geography where the information exchange necessary for innovation *physically* occurs (Kim, 2013). The term 'innovation district' has been loosely applied by the general public to various scales of effort without any industry agreement on the amount of critical mass of people or the resources necessary to define a place as an innovation district (B. a. W. Katz, Julie, 2014). Innovation districts forecasted to cost over $1-2 billion are in the planning stages in St. Louis and Austin (Lower, 2012; Majid, 2014; Marks, 2012).

In 2015, a collaboration began to develop metrics to access innovation districts. The Project for Public Spaces, Mass Economics, Bass Initiatives on Innovation and Placemaking, and The Brookings Institute selected, as a pilot project, Oklahoma City's proposed innovation district adjacent to the University of Oklahoma Health Science complex (B. V. Katz, Jennifer; Wagner, Julie, 2015). Understanding the "critical link between innovation, quality places, and economic growth" is the goal of The Project for Public Spaces. The collaboration intends to produce audit template and tool for innovations districts to adapt for their unique location and goals.

Innovation district examples.

The first innovation district in the world was '22@' launched in 2000 in Barcelona, Spain. The plan was to revitalize a 200-hectare area of mostly abandoned industrial manufacturing to create 150,000 jobs and to attract new companies to the area (A programme of urban, economic and social transformation, 2012). Within ten years the 22@ was home to 7,000 companies (4,500 were new), which employed 90,000 people (56,000 were new). The 22@ area contained 40,000 homes of which 4,600 were built after the year 2000 and another 4,000 were built as government-owned apartments ("A programme of urban, economic and social transformation," 2012).

Barcelona.

The vibrancy and success of Barcelona's 22@ innovation district inspired then Boston mayor Thomas Menino to announce in 2010 the revitalization of Boston's mostly abandoned Seaport District into a 1,000-acre hub for jobs in the 'information age' ("Ciao Innovation District! Menino shares Boston's innovation agenda in Italy," 2010). Within three years of announcement, Boston's innovation district had seen the formation of 200 new companies spawning 4,000 new jobs and over 4,000 residential dwellings. Aside from the new District Hall community building funded by the municipality, most of the development has been from the private sector. By 2014, rents were rivaling the most expensive office space in downtown Boston at about $53 per square foot per year (Ross, 2014). What is not clear in this early example, hailed as a success model, is how many companies and jobs were genuinely new and how many were simply relocated from other parts of the city, state, or country. Measuring true economic gains as opposed to gains from geographical shifts is another analysis entirely.

St. Louis.

The CORTEX District, initially known as the St. Louis Innovation District, is a 200-acre site that links together Washington University, BJC Healthcare, University of Missouri – St. Louis, St. Louis University, and the Missouri Botanical Garden. CORTEX is an acronym for Center of Research, Technology and Entrepreneurial Exchange. The redevelopment plan for the innovation district area indicated the development cost would total $2.1 billion and span from 2013 to 2022 (Lower, 2012). It was projected to be financed by funds from the developers, TIF proceeds (Tax Increment Funds), Transportation Development District, Community Improvement District (CID), Missouri Development Finance Board tax credits, historic tax credits, brownfield tax credits and federal, state and local grants. The City of St. Louis hired a third-party consultant to create a cost-benefit analysis (CBA) on the innovation district plan as it related to additional tax revenues for the City of St. Louis. It projected the estimated assessed valuation (EAV) of the properties would reach $159 million in 2022 from its baseline of $12 million in 2012 (Marks, 2012). The CBA also provided an analysis that estimated 15,639 new jobs would be created with 82% of those jobs having annual salaries of $50,000 to $93,000 (Marks, 2012).

Austin.

The City of Austin describes their innovation district as an *innovation zone* and defines it as:

> A hub of activity dedicated to collaboration, creativity and opportunity; a nexus for exchange of ideas and partnerships, among universities and businesses, a neighborhood to live, work, play and learn within walking distance to transportation – a sense of place; a building providing views of academia and private industry interacting at work, and a catalyst for job creation and economic development. (Majid, 2014)

Geographically, it is a 14.3-acre innovation district planned around the new Dell Medical School and the Dell Seton Medical Center (a teaching hospital associated with the University of Texas). For a complete understanding of their collaborative process approach, review the *Austin Anchors and Innovation Zone: Building Collaborative Capacity* report issued in 2015 (S. R. Greenberg, 2015).

Montreal.

The Quartier de l'innovation (QI) in Montreal, Canada is an interesting example of an evolving university-led innovation district in a large metropolitan area. McGill University (with 39,000 students) and ETS Engineering School (with 8,000 students) saw an innovation district as a way of remaining competitive with other cultural creative cities around the world ("Quartier de l'innovation: a joint vision for a prosperous future," 2014). They cite Toronto, Barcelona, and Boston as their peer cities. Montreal has the highest concentration of post-secondary graduates in North America at 4.38 per 100 people ("Socio-economic Trends - Education," 1996). The second highest is Boston with 4.37 per 100 people. McGill University and ETS Engineering School organized their innovation district by first defining their shared values: to ensure the future economic, social and cultural prosperity of Montreal citizens and community stakeholders (Siles, 2015). They then articulated their goal: to be a creativity city through four pillars: Arts & Culture, Research & Education, Urban, and Industrial (Siles, 2015). They have a 12-person board of directors that coordinates the innovation district projects and development, but they also pull from the collective wisdom of a broader, 24-person steering committee that meets monthly. They value transparency in their operations and are explicit in their purpose: the QI exists to provide experiential learning opportunities for the students, commercialization, and to provide for the betterment of Montreal ("Quartier de l'innovation: a joint vision for a prosperous future," 2014).

Atlanta.

The participation of Georgia Institute of Technology in the 'Tech Square' neighborhood is publically known but scantly documented in the literature (Giuffrida, Clark, & Cross, 2015). The authors state that "although the success of these innovation districts has been widely noted, the elements underlying that success have not been systematically identified" (Giuffrida et al., 2015). Georgia Tech had a traditional campus, but they still took the risk to expand into an adjacent urban area. Their study contributes to this evolving scholarship by examining the development and evolution of Technology Square in Atlanta plus (Giuffrida et al., 2015).

The University

There are three facets of the higher education institutions reviewed for evidence of innovative or utopian thinking. There are: the historical review of the experimental colleges, the innovation campus model currently emerging, and universities known for innovative institutional responses.

Experimental colleges.

Communities of innovators are also found outside private industry clusters or research parks; they exist within higher education institutions themselves. Much of the innovation literature focuses on how the university fosters innovation for *other* products, but there are also great insights available from history that explore how the university innovated its own institution in the quest to create more 'intelligent graduates' (Meiklejohn, 1932). In modern terms, we could call those intelligent graduates *human capital* necessary to contribute to the *knowledge economy*.

After reading a magazine article in which John Meiklejohn elaborated about his ideas for reinventing a new type of college, Glenn Frank, the University of Wisconsin president, invited Meiklejohn to establish a very special type of college within the university: The Experimental College (Meiklejohn, 1932). Aside from how to teach and what to teach, this experiment also investigated the "determining conditions of undergraduate instruction" (Meiklejohn, 1932). At the time, it was felt by those in American colleges that there was a "desperate urgent need" to fuse together the intellectual and social activities of the students (Meiklejohn, 1932).

After the experiment ceased in 1935, Meiklejohn relocated to the University of California at Berkeley where his efforts inspired another wave of experimental colleges within universities from coast to coast. All told, by 1970, over 300 experimental colleges had been started. Ten of that group held a conference in 1964 in Florida to discuss the future of how higher education would continue to innovate as an institution (Stickler, 1964).

The university has long has been a place to experiment with housing to improve the educational experience for students. When Stewart Gordon published *Living and Learning in College* in 1974, he unknowingly was on the cusp of the end of an era in experimental colleges and residential colleges (Shushok Jr, Penven, Stephens, & Keith, 2013). From Gordon's vantage point, he could clearly report the advantages and the pitfalls, but he did not foresee the lull into which the residential college would fall for nearly two decades. The next wave of interest came through various student affairs initiatives in the early 1990s that were prompted by the publication of two key reports: *An American Initiative* and *The Student Learning Imperative* stressed the importance of learning outside the classroom (Shushok Jr et al., 2013). By 2002, 80% of research universities had a *learning community* on campus (Shushok Jr et al., 2013).

There are several terms to identify the innovative restructuring of an educational department or college: innovative colleges, experimental colleges, and distinctive colleges. For ease of discussion, the hundreds of progeny of Meiklejohn's ideas that formed after his educational reform movement until the 1970s are referred to as *innovative campuses* (Kliewer, 1999). As of 1999, there were still 312 experimental colleges in existence (Kliewer, 1999). In groundbreaking research, Kliewer was able to identify a set of common characteristics of these colleges; the one necessity to execute the innovation was the egalitarian approach.

Innovation campuses.

Innovation district developments are typically built on a combination of private property and municipal property in an urban core to leverage the obvious synergies. Often this is near a university campus or involves an educational partner regardless of *proximity*. But some universities are repurposing property and university-owned land to create their own version of an innovation district—often a rural, semi-rural, or suburban interpretation not associated with a dense urban core—and branding it as an *innovation campus*. For a university, an innovation campus—or an innovation district located on or near a campus—is an opportunity to provide experiential learning settings for students and to create new knowledge, new patents, new applications for existing technology, and new opportunities. These developments feature the "intentional co-location of academics and industry to facilitate and streamline the commercialization process" (Bramwell, Hepburn, & Wolfe, 2012). As an evolving hybrid, the innovation campus form can vary, based on existing assets, and they can borrow elements from the research park model and the innovation district form.

North Carolina State University in Raleigh, known as NC State, participates in the RTP, but in the late 1980s, they developed their own unique development 15 miles away. The Centennial Campus is positioned as a model of a public-private *place of innovation* within a 5-minute walk of the main campus. It employs over 11,000 people while 60 industry partnerships maintain a presence on the main campus (L. Tornatzky, 2014). The Centennial Campus at NC State promotional vision video asks "How much in an idea worth? It is the amount of money it makes? Or the innovation it sparks? Or even the lives that it touches?" as a way of reinforcing that knowledge creation and innovation are their academic mission ("Vision 2034," 2015). NC State was identified as one of the top twelve most innovative universities by the authors of the report Innovation U 2.0 (L. Tornatzky, 2014).

In 2015, the University of Nebraska opened its Nebraska Innovation Campus, featuring food innovation as a specialty (V. Miller, Washburn, Norby, Banset, & Klucas, 2015). In Kansas, Wichita State University has incorporated a public maker space in their Experiential Engineering Building in the heart of their "innovation campus" that they are branding as "The Innovation University" (Barrett et al., 2015). Other examples include the Olathe Innovation Campus at Kansas State University, the Missouri Innovation Campus at the University of Central Missouri, the Innovation Campus at South Dakota State, and the Akron Innovation Campus in Ohio (Bramwell et al., 2012). Innovation campuses seem to reside on a continuum between the suburban-inspired research park model and urban innovation district model, but the mere fact research parks are becoming more focused on placemaking gives credence to the influence of innovation district developments. As yet, there is little academic research directed specifically at the innovation campus concept. The degree to which these new *places of innovation* reflect a sustainability orientation is a function of the sustainability commitment of the university and its industry partners; the implications of this orientation – or lack thereof – are discussed in chapter 5.

Innovation as strategic positioning.

In 2002, the *innovation university* concept emerged from thought leaders in higher education in the Research Triangle. In their initial report, *Innovation U*, 12 universities were identified as being the top producers of innovation (L. G. Tornatzky, Waugaman, & Gray, 2002). Since then, many government policies have been implemented that expanded the impact of science, technology, and innovation (STI policies), which impacted the promotion of technology-based economic development (TBED) efforts by the university (L. Tornatzky, 2014). The updated 2012 report, *Innovation U 2.0 Reinventing University Roles in a Knowledge Economy*, profiled the current top 12 universities for innovation and found the following five key areas as vital in order to excel at technological innovation: university culture, durable leadership, entrepreneurship curriculum, interdisciplinary engagement across campus, community, and industry; and engagement in technology transfer (L. Tornatzky, 2014). Only six of the original universities maintained their position on the list, while six others dropped off completely, leaving openings for new innovative universities (Fleming, 2016).

A recent contribution to the conversation about the future of higher education is the book *The Innovative University*; it is credited with introducing the idea of 'disruptive education' to higher education. It provides a glimpse of the drivers behind the impending evolution ahead for higher education institutions and suggests that there are strategies for a university to pursue to innovate, iterate, and evolve into a more resilient institution

(Christensen & Eyring, 2011). For example, Arizona State University was restructured into "twenty-three unique interdisciplinary colleges and schools that, together with departments and research institutes and centers, comprise close-knit but diverse academic communities that are international in scope" (Crow, 2008). Unity College in Maine also reinvented itself with 'Sustainability Science' as the organizing theme and positioned the university as "America's Environmental College," a claim that is validated by recognition from Princeton Review, Sierra Club, and AASHE. For the university to see itself as an agent of change capable of catalyzing innovation, requires it to rethink its internal characteristics including the development of a demand-responsive, creative and collaborative organizational structure that is also flexible and efficient (Allison & Eversole, 2008).

Albert Einstein notes that "the definition of insanity is repeating the same behaviors and expecting a different outcome" so if universities want to retain their position as major contributors of the quickly changing *knowledge economy*, they will have to change behaviors to get a different outcome. Universities should intentionally innovate at the institution level internally and, in that process, learn the skills necessary to promote innovation in external venues. A historical review that can connect how and why universities reinvent themselves, and how forward-thinking universities are reorganizing around sustainability, provides useful perspectives to higher education preparing for and positioning itself for leadership in the *knowledge economy* that fosters *sustainability-oriented innovation.*

Innovation Ecosystems and Entrepreneurial Ecosystems

Industrial districts, clusters, research parks, innovation districts, and universities are physical places where systems operate to support the ventures of innovators and entrepreneurs. These *places of innovation* are not just geographical places but rather venues for networks and systems popularly referred to as *innovation ecosystems* and *entrepreneurial ecosystems*. These 'ecosystems' warrant defining because of their ties to fostering innovation.

The term *entrepreneurial ecosystem* was introduced as a working theory by Jude Valdez, a University of Texas at San Antonio professor, presenting at the Small Business Institute Director's Association Conference in 1988 (Valdez, 2000). The term became popularized in the academic literature, then found an advocate in Daniel Isenberg of Babson College who published a series of articles about *entrepreneurial ecosystems* in the Harvard Business Review (Isenberg, 2011). Isenberg's contributions are significant because Babson College has the top-rated entrepreneurship division in the United States and maintains a satellite office in the Seaport Innovation District in Boston. The term *entrepreneurial ecosystem* broadly describes the resources and actors needed for an entrepreneur to thrive. The *entrepreneurial ecosystem* is deemed to have merit as a "metaphorical device which offers a holistic understanding" and allows the firm's growth to be considered as a function of the external environment rather than solely considered as a function of its internal characteristics (Mason & Brown, 2014). Research on case studies identified the seven key components necessary to create a university *entrepreneurial ecosystem* (Rice, Fetters, & Greene, 2014). Brad Feld, a practitioner and venture capitalist, published *Startup Communities*: *Building an Entrepreneurial Ecosystem in Your City* that describes how Boulder, Colorado transformed itself into an *entrepreneurial ecosystem* over a twenty year span (Feld, 2012).

The idea of an *entrepreneurial ecosystem* uses biology to draw parallels, but then the term was expanded to create a new term 'entrepreneurial ecology,' which is designed to infuse sustainability into entrepreneurship (Frederick, 2015). A similar concept, 'sustainable entrepreneurial systems,' was set forth as a guiding framework for Victoria Island in Canada to steer innovation toward a sustainable society (B. Cohen, 2006). In a parallel development, *sustainability-oriented innovation* has emerged over the past decade out of the eco-innovation research premise (Klewitz & Hansen, 2014). Eco-innovation is defined as: "all measures of relevant actors which: develop new ideas, behavior, products, and processes, apply or introduce them and which contribute to a reduction of environmental burdens or to ecologically specified sustainability targets" (Klemmer, Lehr, & Loebbe, 1999). Within the general innovation literature field there is a thread of research developing around sustainability goals.

The mindset for an *entrepreneurial ecosystem* became the driving influence for restructuring Arizona State University into the New American University, a new model of higher education for the twenty-first century according to President Michael Crow. He is building an *entrepreneurial university* based on the belief that the primary asset of every college is the fusion of intellectual capital with creativity and innovation (Crow, 2008). The goal of ASU is to "redefine public higher education through the creation of a prototype, solution-focused institution that combines the highest level of academic excellence, maximum societal impact, and inclusiveness to as broad a demographic as possible" (Crow, 2009). In 2006, ASU also built a 1.5 million square foot facility called SkySong Center to serve as an innovation center. ASU provides an example of what an institution evolves toward with the application of an *entrepreneurial ecosystem* overlay.

Innovation ecosystems are related to *entrepreneurial ecosystems* and even co-exist in the same realm; the former is designed to accelerate innovation while the latter is intended to promote the growth of a venture (Morrison, 2013). One paper attempted to wade through the literature's "loose and inconsistent use of the term" to build an understanding of how an *innovation ecosystem* differs from a science park, innovation cluster, and regional innovation systems (Oh, Phillips, Park, & Lee, 2016). Ed Morrison, at Purdue University, is a hybrid academic/practitioner who catalyzes the innovation potential of a place in higher education institutions through a 'collaborative strategic doing' framework that leverages brain power, *entrepreneurial ecosystems*, *innovation ecosystems*, and quality places (Morrison, 2014).

One of the most successful *innovation ecosystems* in the world is Silicon Valley. Because it is revered as a model of unparalleled success for technological innovations and scalable entrepreneurial start-ups, the intangible mechanisms and the physical attributes have been extensively studied. Historical context also explains how the Valley was uniquely formed by the unplanned confluence of Cold War spending, GPD growth, immigration, risk-tolerant capital, entrepreneurial leadership and good weather (M. O'mara, 2011). Cultural historians Leslie and Kargon have studied the Valley extensively and concluded there were too many unique aspects and unknown mechanisms in the Valley to hope to replicate its *innovation ecosystem* in another place (Leslie & Kargon, 1996). Nevertheless, the mystery did not stop researchers from trying to unlock the secrets of the success of Silicon Valley.

The university participation in Silicon Valley began when the Stanford Research Park was built in a 200-acre field at the edge of Palo Alto, California in 1951 (Sandelin, 2004). Silicon Valley epitomizes ideas generated from *proximity* of talent (Engel & del-Palacio, 2011). The abundance of concentrated talent results in spillover of knowledge that benefits firms in the area (Audretsch & Keilbach, 2007; Maxwell, 1992). The Valley is the gold standard of innovation-spawning networks; much is written about how to replicate their success (Hwang & Horowitt, 2012; M. P. O'Mara, 2015). Even a research study that benchmarked the top 10 university *entrepreneurial ecosystems* in the world concluded that those hoping to replicate the success of MIT and Stanford "would be much better off studying their early history than trying to copy what they are doing now" (Graham, 2013). Etzkowitz echoes this sentiment toward studying history by stating "focusing the on the visible manifestations of Silicon Valley's ecosystem, its incubators, angel networks, and venture capital firms may only transfer a façade, neglecting the creation of the *Triple Helix* foundation of university-industry-government interactions on which the highly visible ecosystem rests" (Etzkowitz, 2013). The field of economic geography has generated wide agreement that ideas for products and processes are created by collaborative networks of people and are accomplished via the *proximity* of place (Chatterji et al., 2013).

One reason offered for the plethora of innovations originating in a specific place is the 'knowledge spillover' effect that occurs when *proximity* and informal social interaction (often in the public realm) leads to the formation of trusted social networks that contain an excess of knowledge. This 'abundance of knowledge' overflows to other people who have a use for it (Audretsch & Keilbach, 2007). People who engage in conversations with each other sporadically, but not frequently, are considered *weak ties* because the information they share tends to be diverse due to the wide variety of social circles involved; ironically, there is great strength in *weak ties* (Granovetter, 1983). *Strong ties* suggest that people in the same social circles will produce duplicate information, whereas *weak ties* have the potential to generate a wider variety of unique information because of its diversity and connectivity to others.

Aside from Silicon Valley, the technology corridor on Route 128 in Massachusetts and the Denver-Boulder area have also emerged as hubs of innovative entrepreneurship that are engaged with university collaborations (Galbraith, 2012). Using such large and established *innovation ecosystems* as models has limited application for smaller cities or newly established ecosystems. For mid-sized legacy cities, such as Detroit, the resources are not available to saturate an entire region with innovation, so their energy is better utilized at the city scale thus the burgeoning interest in innovation districts (Flint, 2016).

When a large regional *innovation ecosystem* is not feasible or necessary, an innovation district development can be employed. The innovation district has been described as a "science park at city-scale" (Price & Delbridge, 2015). In other words, the innovation district is bigger than a science park and smaller than a regional *innovation ecosystem*.

To increase the likelihood of innovation flourishing and leading to economic growth, a variety of metrics to measure their creativity and innovation were developed by academicians and practitioners (R. Florida, 2014; R. L. Florida, 2002; Hartley, Potts, & MacDonald, 2012; Landry, 2005; Ponzini & Rossi, 2010). Since 1995, Silicon Valley has maintained their own custom index to measure annually their "bellwethers that reflect fundamentals of long-term regional health" (Network, 2004). Victor Hwang, a former Silicon Valley thought leader on innovation and author of the Rainforest Scorecard, developed a methodology to measure corporate innovation for the purpose of understanding the leverage points; he describes their work as "architects of the invisible" (Hwang & Horowitt, 2012). Though these levers may be invisible, they are mechanisms used to create the conditions that foster innovation.

What is often discussed is how Silicon Valley can be replicated and what constitutes an effective *innovation ecosystem*. What is missing from the conversation is what process operates inside those environments and how those processes might be applied to other innovation venues: smaller cities, rural communities, neighborhood efforts, and higher education communities. It is a misconception that only the largest, most densely populated cities can innovate (Orlando & Verba, 2005). Orlando and Verba find "even in

96

remote locations, researchers can acquire knowledge from others if they know exactly whom to contact" and they find this is especially true for mature industries that continue to innovate (Orlando & Verba, 2005).

The quest for these various metrics is to find a set of replicable features that foster innovation. Yet, the 'feature' is not what produces the innovation; it is the 'process' that generates innovation. As an analogy, the 'processes' are like the action verbs between the nouns of the *innovation ecosystem* features. How innovation is spawned from within these ecosystems is an emerging field that can advance understanding about replicating *sustainability-oriented innovation* in the future. Missing from the wider innovation discourse and the ecosystem metrics is any kind of discussion about the evaluation of the overall sustainability impact of the innovations, or whether these hotly pursued innovations are *sustainability-oriented innovations* in support of a different kind of economy that reconciles itself to environmental constraints. With the 2015 launch of the Eco District protocol that is overlaid on an innovation district (or an ecoinnovation district) those pilot projects may create a rich area of research on this exact topic (J. Miller, 2015).

The research gap this dissertation fills is the perspective created by viewing these *places of innovation* as a whole: industrial districts, clusters, research parks, innovation districts, and universities and analyzing them collectively to provide higher education thought leaders with a history of how people in those places approached their charge to innovate. But before a university can engage meaningfully in innovation external to the campus, or a co-location collaboration with industry partners on a campus, it must first appreciate its

own innovations within the history of higher education through a discussion on

experimental colleges and innovation universities.

CHAPTER III

METHODOLOGY

The Qualitative Interpretations for Quantitative Outcomes

Qualitative interpretations are necessary to contextualize quantitative outcomes because qualitative interpretations answer *how* the quantitative outcomes came to be. Quantitative examples of research on innovation outcomes and outputs proliferate throughout academic journals that publish on economic geography; these examples are vital because they establish that these *places of innovation* bring tangible economic value. Quantitative research demonstrates that *places of innovation* provide a satisfactory return-on-investment that warrants the expenditures of time, energy, and money used to create these places. Without that reliable ROI, investing money in *places of innovation* would be based on a more intangible, more difficult to justify rationale. However, the quantitative research does not and cannot offer satisfactory explanations of how these places became successful economic engines; simply knowing that these places are successful does not explain why they are successful. In order to understand the mechanisms that generate successful *places of innovation*, we must rely on qualitative investigations. Qualitative investigations reveal the non-quantifiable mechanisms at play.

Researchers have contrasted the value of patents in a non-clustered area with the value of patents generated within the 22@ Barcelona innovation district (Boix & Galletto, 2009). Seminal contributions have also come from Michael Porter at the Harvard Business School when he published on the economic value of clusters of like industries and competition (Porter, 2008a). These monetary return valuations are necessary and useful measures of output but do little to describe the inputs (or precipitants) that produced innovation value or the characteristics of the place that fostered those successes. The qualitative methodology used in this dissertation allows for theory-generation from data about these inputs found in the published narrative accounts in the histories of *places of innovation* and the histories of *intentional communities*. Narratives were found in hundreds of printed sources including historical books, academic books, popular press books, newspapers, association websites, industry trade publications, and peer-reviewed journals. This review of sources yielded historical analogies capable of generating correlations between persistent themes found in communities and innovation. Research empirically establishes that *places of innovation* can effectively produce desired results, so now research attention can be turned to a causal investigation about *how* and *why* these places work and can seek to understand the replication potential for higher education institutions.

The Role of the Researcher

Often times it is through a researcher's background, ethics, and motivations that a particular deficit in knowledge is recognized. What unique insights does the researcher know? What does the researcher stand for? Why does the researcher care? Because each researcher is unique in his or her worldview, each possesses the ability to see gaps in research that others miss. Understanding the background of the researcher who designed and executed the study provides insight into the qualitative validity. Within the social science fields that employ narrative inquiry, it is understood that the motivation of an author's experience, expertise, and particular life world drives the research toward a personally meaningful line of inquiry (Grant, 2010). Because the learning curve to understand different contexts is steep, the researcher must have abundant experience in these contexts to be qualified to make valid inferences (Steinberg, 2015).

In the decade-long quest to understand sustainability in theory and in practice, I read about it, studied it, followed it, wrote about it, experimented with it, personally applied it, taught it, and discussed it with thousands of people. To explore *intentional communities*, my travels took me to The Farm ecovillage in Tennessee and the academic ecovillage at Berea College in Kentucky. To explore *places of innovation*, my travels took me to innovation districts in Chattanooga, Tennessee; Milwaukee, Wisconsin; and Seattle, Washington and to an 'innovation university' at Wichita State University in Kansas. To explore sustainability deeply, my travels took me nationwide to dozens of practitioners' conferences and higher education sustainability conferences. This field research did not capture or record data, but rather it fueled the skill of reading the traits of the industries, is known in academic terminology as "empathic accuracy" (Ickes, 1993). This informal field research also provided experiences to build expertise vital to making inferences in future research analysis (Collier, 2011). As former Apple CEO Steve Jobs was fond of explaining about his journey that cultivated his innovative mind, "of course it was impossible to connect the dots looking forward when I was in college. But it was very, very clear looking backwards 10 years later" (Jobs, 2005). It is a rare opportunity to be in the midst of a historical turning point and in the position to chronicle impressions of it through academic research. Wearing the dual hats of innovative entrepreneurship and environmental science provided a more holistic interpretation than would have been impossible if experienced from a single discipline perspective.

As an environmental scientist who has worked as a researcher and writer in the green building, architecture, planning, and sustainability industries and who also worked as a university instructor in business sustainability and sustainable communities, I have developed a unique perspective on higher education's potential to use place and community – beyond the conventional classroom – to prepare citizens to be capable of innovating solutions toward the creation of a sustainable society. Specifically, the fields in which I worked provided evidence of innovations around sustainability that gives credence to the gaps in knowledge this research addresses. Those fields exposed me to green building and sustainability metrics for cities in the built environment and curriculum development for sustainable business and sustainable communities in higher education; all of which were experienced through the lens of environmental science.

The environmental science lens.

A perspective grounded in environmental science set the foundation to interpret the changes that have happened over the past few decades. The *Blueprint for Survival* publication, authored by Edward Goldsmith, was a call for radical change through political organizing to address pressing global environmental issues (E. Goldsmith, Allen, Allaby, Davoll, & Lawrence, 1972). The popular definition of sustainable development emerged from a report entitled *Our Common Future* published by the World Commission on Environment and Development (also known as Brundtland Commission) in 1987: "development which meets the needs of current generations without compromising the ability of future generations to meet their own needs" (Brundtland et al., 1987). An even more esoteric but honest interpretation of sustainability that is appropriate for this research comes from Cary: "sustainability is not a fixed ideal, but an evolutionary process of improving the management of systems, through improved understand and knowledge. Analogous to Darwin's species evolution, the process is non-deterministic with the end points not known in advance" (Cary, 1998).

A global conference on sustainable development in 1992 hosted by The United Nations Conference on the Environment and Development (informally known as The Earth Summit and often referred to simply as "Rio") produced a global call for action in the form of three documents: Rio Declaration, Agenda 21, and The Forest Principles. While these documents achieved consensus, other global thought leaders thought the time was coming when a more strongly worded document was necessary. Subsequently, the Earth Charter introduced in 2000 was "an ethical framework for building a just, sustainable, and peaceful global society in the 21st century" (Initiative, 2000). The 2,400-word document featured in Table 1 in full is revered as the most concise framework for

sustainability; it sets forth 16 principles focused in the areas of community, ecology,

justice, and democracy (Initiative, 2000).

Table 1 – Earth Charter Values Declaration

EARTH CHARTER 2000
Respect and Care for Community Life: • Respect Earth and life in all its diversity.   • Care for the community of life with understanding, compassion, and love.   • Build democratic societies that are just, participatory, sustainable, and peaceful.   • Secure Earth's bounty and beauty for present and future generations.
Ecological Integrity: • Protect and restore the integrity of Earth's ecological systems, with special concern for biological diversity and the natural processes that sustain life.   • Prevent harm as the best method of environmental protection and, when knowledge is limited, apply a precautionary approach.   • Adopt patterns of production, consumption, and reproduction that safeguard Earth's regenerative capacities, human rights, and community well-being.   • Advance the study of ecological sustainability and promote the open exchange and wide application of the knowledge acquired.
Social and Economic Justice: • Eradicate poverty as an ethical, social, and environmental imperative.   • Ensure that economic activities and institutions at all levels promote human development in an equitable and sustainable manner.   • Affirm gender equality and equity as prerequisites to sustainable development and ensure universal access to education, health care, and economic opportunity.   • Uphold the right of all, without discrimination, to a natural and social environment supportive of human dignity, bodily health, and spiritual wellbeing, with special attention to the rights of indigenous peoples and minorities.
Democracy, Nonviolence, and Peace: • Strengthen democratic institutions at all levels, and provide transparency and accountability in governance, inclusive participation in decision making, and access to justice.   • Integrate into formal education and life-long learning the knowledge, values, and skills needed for a sustainable way of life.   • Treat all living beings with respect and consideration.   • Promote a culture of tolerance, nonviolence, and peace.

In 2001, the United Nations began collaborating with 1,300 experts on another set of assessment reports that chronicled the changes in the environment due to human impact. The Millennium Ecosystem Assessment, released in 2005, suggested substantial policy changes would be necessary to reverse the damage to the ecosystems and protect the human lives. More recently, the Stockholm Resiliency Center produced an interdisciplinary report identifying the nine ecological boundaries necessary to balance in order to avert global ecosystem collapse; four of the nine boundaries have been destabilized by human impact (Rockström, 2009). The book *Peak Everything* published by Richard Heinberg in 2010 listed the major natural resource areas prone to collapse (water, fossil fuels, climate change, etc.) and addressed what changes would be necessary to prevent a global societal collapse.

This backdrop of environmental destabilization illustrates the dire need for innovations around sustainability. There is a link between finite resources and the future sustainable innovations yet to be invented that will restore balance and sustain civilization. The conventional business researcher likely does not have the bias from an environmental education that an environmental scientist embodies, so disclosing this academic foundation in methodology is necessary to appreciate the vantage point of the lens of the research, the analysis, and particularly the implications. Chapter 5 establishes a foundation for implications by providing a concise historical summary of the recent responses to innovations around sustainability that have occurred in the greening of the built environment, the greening of business, and the greening of higher education.

The Data

The evidence originated in two general fields: *intentional communities* and *places of innovation*. In some respects, these fields are juxtaposed in a way that offers the potential for creative insights useful in novel theory building (Eisenhardt, 1989). Upon deeper reflection, they are also very similar in that they are both *communities of innovators*. Over 1,000 printed sources including historical books, academic books, popular press books, the occasional newspaper article, association websites, industry trade publications, and peer-reviewed journal articles were digested to develop broad contextual understanding and then they were scoured for commonalities. The topical areas for *intentional communities* found historical evidence that identified the characteristics that cultivated innovation in utopian studies, *intentional communities*, cohousing, ecovillages, academic ecovillages, and innovation community prototypes. The topical areas around *places of innovation* were industrial districts, clusters, research parks, innovation districts, and universities. The number of sources used for building a context, based on articles in the Endnote referencing software, are listed in Table 2.

Table 2. Sources Used as Evidence

Topical Areas	Books	Articles	Websites
Intentional Communities	2	219	2
Utopian Studies	7	132	-
Innovation Prototype Community	2	6	2
Cohousing	1	42	2
Ecovillages	9	108	15
Industrial districts	2	22	1
Clusters	2	27	1
Research Parks	1	12	1
Innovation & Entre. Ecosystems	-	23	-
Green Business	6	30	5
Open source / Open innovation	2	22	1
Higher Education	8	159	10
Sustainability & Environmental Science	88	197	14
Architecture & Planning	31	35	12

The logic for including so many topical areas in the analysis is that they were evolutionarily related and interconnected. Also, diversity of data sources increases the generalizability of the results (Steinberg, 2015). Open-ended research questions have a tendency to produce staggering amounts of data for a within-system investigation, but that also allows for unique patterns to emerge without the pressure to generalize a pattern across all the evidence (Eisenhardt, 1989).

When performing electronic library database searches for evidence, the keywords used were the topical areas listed in the chart. Also, the references listed in the academic articles became springboards for additional evidence gathering. The primary database used was Google Scholar. Looking for patterns led to the formation of themes; these were coded as 'OS' for *open source*, 'IP' for *iterative process*, or 'P' for *proximity* in the margins and by applying color-coded sticky notes to the hard copies. When evidence was found contrary to the themes explored, that too was earmarked for later evaluation in the process tracing analysis. A single source could generate multiple points of evidence that contributed to the emerging theory; the estimated N is 200 from 1,200 potential source documents. The research goal was to identify correlations without undue redundancy.

The analysis of evidence served two purposes: theory-generating for this dissertation with the potential for theory testing in future research. In consideration of long-term applied research, the intent was to identify an emerging theory with the ultimate goal of making it testable and transferable. The early passes through the sources produced reoccurring concepts that coalesced into themes and those themes constituted an emerging theory. Subsequent analysis through the sources also outlined the basis for cross-system generalization for theory-testing for future research discussed in chapter 5; this was specifically tailored for application to emerging innovation districts and higher education institutions.

Interestingly, it was the commonality contained in these topical areas that revealed the overlaps that developed into the themes used in theory building. The more observations of different kinds creates a theory that is likely more valid than a theory developed from limited observations, the goal being to develop "explanatory variables" rather than just simple correlations (Hall, 2006). The two topical fields offered a dozen subfields to review for observations. By increasing the number and diversity of the literature sources, the investigator's confidence is increased that the causal process is not just idiosyncratic (Hall, 2006). What matters more than the sheer number of observations though is the strength of the relationship that is established between the evidence and the hypothesis driving the theory (Bennett, 2010).

The data narratives were derived from sources that were gathered between 2005 and 2016; sources included academic literature, books, and industry publications. The actual publications were from authors of the twentieth century and twenty-first century about historical topics; one source was an exact reproduction of the book *Utopia* as originally published in 1516. Following the initial search that began in 2005, several years were spent creating context from industry experiences in green building and reading widely in sustainability. The ecovillage focus began in 2010 as an offshoot of sustainable cities and sustainable communities research used in teaching. Green entrepreneurship became an offshoot of ecovillage economies. Local economic development based on innovation placemaking came to my attention in 2014, when a practitioner explained how innovation districts were in the formation stages.

To me, the research investigations seem to unfold logically as if connected and nested within a bigger picture. As John Muir (1838-1914) says, "When we try to pick out anything by itself, we find it hitched to everything else in the Universe" (Muir, Limbaugh, & Lewis, 1985). Each 'rabbit trail' leading off the main investigation was a nutrient-seeking root tied to the taproot that nourished the tree of knowledge. Likely, sustainability scholars are comfortable with ambiguity and ridiculous amounts of connectedness. The final data analysis focused on historical trends to explain the present evolution happening around communities of innovators in specific types of places. The themes that were ultimately identified did so by congealing slowly through years of patient iterations that involved absorbing broad patterns, noticing commonalities, and

grounding them in the greater context of sustainability.

Historical Methodology

Methodology is the study of approaches of producing knowledge through research. Method refers to a specific procedure used to collect the data that will be analyzed to create that knowledge. Within any given methodology there are methods available to deploy, depending on the research design, type of evidence and data sought. The methodology used is historical methodology and the method utilized is process tracing; both are discussed in detail in the following section

Within historical research methodology, an approach called narrative inquiry is commonly used to identify and interpret the data collected (D. J. Clandinin & Connelly, 2000). Narrative inquiry can be executed by using a systematic procedure referred to as process tracing, originating from the academic field of comparative politics. Infusing the historical method with environmental scanning from the practice of scenario planning allows researchers not only to evaluate the history to the present time, but also to evaluate the trajectory of the historical trends (Staley, 2002).

Theoretical Framework: Scientific Realism

Scientific realism is based on the premise that a scientific theory can contribute to knowledge about the unobservable and that an approximate truth can be an adequate explanation because it progresses toward the true nature of the physical world (Psillos, 2000). Scientific realists view theories as the journey to discover the truth and, as such, are incremental steps toward knowing. Process tracing proponents are clear that causality

is not observable, but inferences about causality can be made (Bennett & Checkel, 2014).

The History of the Historical Method

Presenting historical research with an embedded scientific approach may seem obvious to the modern researcher, but there was a time in the development of history as a science when the scientific method for use in history was a matter of great debate. A methodological revolution in history was made possible through the standards set by Leopold von Ranke in the nineteenth century and the positioning by Reinhold Niebuhr in the twentieth century that allowed the "independent science of history to become a true intellectual discipline" (Sreedharan, 2007). History is studied so causal explanations can bring understanding to static correlations and those can then be developed for generalizations in other contexts in the present and foreseeable future. A generalization is "a logical argument for extending one's claims beyond the data" (Steinberg, 2015).

To achieve "historically specific" explanations around a particular event, special attention is applied to "preconditions, precipitants, and triggers" and how these interact in context to generate an outcome (Hall, 2006). The events subject to analysis are the many narratives and their observations about *intentional communities* and about *places of innovation*. A *theory-oriented explanation* seeks to identify the most important elements in the causal chain by generating inferences about an outcome based on a multiplicity of observations, existing studies, and intuition (Hall, 2006). One issue, though, is the "facts against which a theory is tested are always generated, to come extent, by the theory itself" and this discernment of facts calls for a "fine-grained judgment" (Hall, 2006). After establishing simple correlations between explanatory variables of an outcome, the case for causality should include a rationale for *why* (Hall, 2006). Observation offers correlation; causality offers explanation, but the underlying root cause is only understood through the patient exploration of *why* and the degree to which the *why* is deemed credible becomes the basis for accepting the generalizations offered. Justifying the *why* creates the compelling logic of generalization. When the *why* explains the relationship it also establishes internal validity (Eisenhardt, 1989).

The other practice to increase the confidence of the investigator's data is to increase the number and diversity of cases. Given that this dissertation explores evidence within dozens of subcategories within the topical area, the inferences should be considered reflective of the theory they build. Specifically, because the data originated in many complex systems involving different people living in different times and different locales and holding different motives, the historical narratives from comparative political inquiry promise that theory can be derived from phenomena in different settings (Steinberg, 2015).

Mark Twain might agree, historical analogies are not exactly history repeating itself but they could be history humming a familiar rhyme. Although the historical analogies drawn are not predictive, they do reveal the possible rather than the probable and the anticipatory rather than the predictive; historical analogies allow for preparation rather than control (Gottschalk, 1966). Preparation, anticipation, and identifying possibilities are reasons historical methodology is very suitable for discourse about how to prepare for the future.

The historical method is more than just facts in chronological order; in fact, in a seminal paper written over one hundred years ago, Robert Hoxie chastised the concept that history should be just a chronological narrative (Hoxie, 1906). He passionately argued that an irrelevant narrative, devoid of meaning, paled against the tremendously useful application of a genuine historical method. He asserted that the historical method was a valid scientific method because it starts by stating a hypothesis and gathering data by strategically selecting events in history for the purpose of supporting (or refuting) the hypothesis (Hoxie, 1906). This sentiment continues to be echoed by modern historians a century later (Clayton, 1996).

Gottschalk's Five Principles

Given the nature of the research questions of this dissertation, a qualitative method was more appropriate than a quantitative method, so Gottschalk's five principles of historical research were applied as a critical assessment of its benefits and disadvantages inherent in the method *as well as in the researcher* (Gottschalk, 1966). Justification of the historical methodology starts with not just listing Gottschalk's five principles but exploring the implications of each sentence, in context, to the research for this dissertation. Therefore, the principles are initially summarized in full below and then dissected individually in five parts to allow for a fuller consideration in advance of application. According to Gottschalk:

> After the hypothesis is articulated, first collect data. Gather all that is considered relevant in justifying a judgment. Second, conscientiously give full weight to the data but no more and no less than it deserves. Third, strive to eliminate biases regardless of how unreasonable that is to achieve. Fourth, suspend judgment in the absence of definitive testimony. Lastly, avoid gratuitous assumptions and inferences, instead opting to make conclusions that logically proceeded from the evidence. (Gottschalk, 1966)

We begin using Gottschalk's framework by evaluating the first of the five parts:

> After the hypothesis is articulated, first collect data. Gather all that is considered relevant in justifying a judgment. (Gottschalk, 1966)

Every historian has two points to establish with the documents referenced: are they authentic and are they credible? (Gottschalk, 1966). Unlike an experiment in a lab, the past offers millions of data points in time from which to choose in a historical analysis. There must be a quality of discernment present in the researcher to know where to look, what to collect, what to cull, and when to stop. Initially, much evidence seems relevant, few pieces actually are; redundancy becomes apparent. This overabundance of evidence is especially characteristic for inquiry-based research.

Additionally, historical data are not generated for specific scientific purposes. They were compiled by an array of writers of various quality and training for unscientific purposes; many historical narratives are not captured with the intention they would be used in an academic sense. Therefore, while abundantly available, the data are incomplete and meaningless without analysis, interpretation, and context. Analysis and interpretation are matters of fine-grained judgment, exercised with vigilance and curiosity and intentionally correcting itself of bias in pursuit of genuine scientific exploration. Many times, this means reading with fresh eyes the same article multiple times to glean new insights.

There can be a tendency in human nature to over-generalize and cherry-pick the data to justify a predetermined outcome. Cherry-picking is the intentional suppression of evidence that counters a sought-after theoretical argument. It is disingenuous of an academic scholar who takes seriously the task of advancing knowledge through contribution. To counter this, all evidence must be included in findings. This includes counter evidence and a lack of evidence. Research that fails to establish a causal relationship is viewed as inferior and is less likely to be submitted for publication. This publication bias can drive a researcher to make sweeping generalizations that are detrimental to quality historical research (Sreedharan, 2007). Generalizations are easy and statements of quick judgment usually reflect an intellectually shallow approach that lacks respect for holistic perspectives.

To establish relevancy requires a deliberate process of critically examining validity of the sources. The term *process* implies it is not a single-session decision; often it takes a preponderance of evidence to accept or reject historical fact. What expertise can be used to access the truthfulness or intentionality of the evidence? Each recorded piece of evidence is riddled with hidden motives unknowable to the researcher evaluating it. The process of critically examining sources can be considered an endless exploration. Carr's famous characterization of history defines it as "the unending dialogue between present and past" (Carr & Davies, 1961).

Moreover, historical data contains information gaps or errors (Simonton, 2003). Given that the sources of evidence for this research start in British literature in 1516 and go through 2016, the wide time range suggests the likelihood that there are errors of omission in the early years and information inundation in the latter years. The historical accounts of *intentional communities* in the eighteenth and nineteenth centuries were not captured and published as books until twentieth century authors conducted historical research on them. Several of the twentieth century books reviewed were published by former members of *intentional communities* and reflect their personal biases as much as a historical representation of the community. Additional gaps exist because some innovations are so new there is a time lag before the topic is published in academic literature. The data choices range from limited choices to an abundance of choices and from ancient sources to the very recent.

From a sampling standpoint, the data gathered was not random, though it was selected from a very broad collection of evidence in excess of 900 publications. I used my judgment and experience to select what was historically significant and representative of *intentional communities* and *places of innovation*. While removing subjectivity is an admirable goal of qualitative research, it is virtually impossible because with subjectively comes the evaluative judgment necessary to write historical accounts. The motive must be clear though; the research agenda is one of advancing the field of knowledge and not the ideological stance of a subjective researcher with a hidden agenda. I searched for interview transcripts embedded in scholarly publications so the robust narratives quoted would originate from 'thick descriptions' rooted appropriately in context (Ponterotto, 2006). Some of this data collected were cited from primary sources; the narratives replicated in the publications that were spoken or written by people who were firsthand witnesses, which give great validity to the evidence investigated (Gottschalk, 1966).

Second, conscientiously give full weight to the data but no more and no less than it deserves. (Gottschalk, 1966)

There are many reasons why data would not be measured fairly, either by accident or by intent. If there are ulterior motives or unacknowledged biases, then the interpretations could vary from one extreme to another. Knowing how much weight to give a piece of evidence requires discernment. Even if hundreds of pieces of data seem to indicate a conclusion, the quantity of evidence is less important that the quality. One quality piece of evidence can outweigh one hundred suppositions. Recognizing the appropriate weight to give each piece of historical evidence is a skill sharpened by experience. Philosophical principles are derived from life experiences and it is these principles that create the ability for judgment, which is a required criterion for good descriptions and good data selection (Gottschalk, 1966).

Third, strive to eliminate biases regardless of how unreasonable that is to achieve. (Gottschalk, 1966)

Aside from the importance of validity, understanding is an even more fundamental concept for solid qualitative research (Wolcott, 1990). Understanding is affected by how a researcher perceives reality through the filters of culture, language, values, beliefs, attitudes, expectations, and intentions.

The vetting process of authenticity and credibility is accomplished with an intuition or empathy referred to as "historical open-mindedness" thus positioning the researcher to

121

shed his own personality and ethics and review the data in context (Gottschalk, 1966). Biases lead to intellectual distortion (Sreedharan, 2007). Biases also render objectivity impossible (Sreedharan, 2007). This is countered with an excess of evidence that it cannot be consciously denied. Since biases are inherent in all human beings, a researcher can acknowledge them and strive to transcend those restraints so that a narrative can be rendered based on other people's multiple perspectives in context. Utilizing a wide variety of sources is also useful in increasing objectivity to overcome prejudices (Sreedharan, 2007). Genuine open-mindedness is a rare quality in general and even more so in research; it requires humbleness, healthy skepticism, and comfort with ambiguity.

> Fourth, suspend judgment in the absence of definitive testimony.
> (Gottschalk, 1966)

At best, the evidence from historical sources is considered weak because it does not meet the criteria of internal validity (Simonton, 2003). A researcher needs to be very careful when suggesting a causal link but, nevertheless, that link is why historians exist. When the field of history was evolving, a school of thought emerged that historians should avoid the why questions and just record how events unfolded and avoid interpretive explanations. Historian E.H. Carr took a strong position against that school of thought and advocated that "the study of history is a study of causes" (Carr & Davies, 1961).

Only when definitive testimony is available can a historian render a judgment. The researcher has an obligation to postulate a reason why events occurred and to make some meaning from them so that it can bring value to understanding the present situation at hand and be informative regarding future actions.

Lastly, avoid gratuitous assumptions and inferences, instead opting to make conclusions that logically proceeded from the evidence. (Gottschalk, 1966)

Historical researchers are called to draw conclusions but not to make unwarranted assumptions. The reason a researcher might make an unfounded inference would be a desperate attempt to validate their conclusions. Like a judge, a researcher needs to consider all evidence with no prejudice and then carefully, methodically, and ethically come to a logical conclusion that others can trust and accept.

The power of the narrative is strong and subject to hijacking by imaginative writers who might be passionate about solidifying their positions. Nineteenth century historian Fred Fling wisely cautioned that "the uncontrolled imagination is a dangerous thing in history and leads to false conceptions and combinations" (Fling, 1899). Curiosity and imagination are vital tools to mine the past for data but, left unchecked, they can be the exact qualities that can destroy academic credibility.

Again, the researcher must apply passion to the topic for the joy of the discovery and not seek overtly to win over the hearts and minds of the reader for a predetermined purpose. Arguments put forth in academic contribution should be convincing due to their logic, not their lawmanship (Gottschalk, 1966).

Gottschalk refers to these five principles as embodying the "scientific spirit" because, as imperfect human beings, these are lofty goals to which we can only aspire but never fully reach. Only by acknowledging the difficulty of each of the five principles do they become stronger principles and thus increase the expertise of the researcher.

The narrative inquiry framework is what allows the researcher to follow and fulfill Gottschalk's principles by identifying the common threads of the data selected and weaving together an explanation of the present that is useful to inform the future. Within the narrative inquiry framework, there are three paths of justification: personal, practical, and social (D. Clandinin, Huber, McGaw, Baker, & Peterson, 2010). It is in the social justification realm where the questions "so what?" and "who cares?" can drive the inquiry and thus provide theoretical contributions to the discipline (D. Clandinin et al., 2010).

Process Tracing Method: Inductive Investigations Begin with Experience

Social science research is considered iterative and inductive (R. Bates, Greif, Levi, Rosenthal, & Weingast, 2000). The research around this dissertation can be accurately described as "unintentionally inductive" because from day one at the university, I sought for an explanation of *how* a community came to be a certain way, *how* a place produced innovators and *how* higher education could recognize and leverage those qualities to itself to achieve its sustainability mission for the benefit of humanity.

This circling of these inductive questions resulted in deeper and more concise iterations. Ultimately, the realization that achieving sustainability would require *communities of innovators* led to the convergence of all the fields of study. The collection of data was also driven by instinct from a collection of life experiences. This "iteration between theory and data" has been recommended as a viable way to transform an approach to problems into a research area (R. Bates et al., 2000; Eisenhardt, 1989). Eventually, the vast amounts of preliminary observations served to create very informed and meaningful research questions capable of application to the emerging area of innovation for sustainability in higher education.

The Systematic Procedure

A basic definition for process tracing is "the use of evidence from within a historical case to make inferences about causal explanations" (Bennett & Checkel, 2012). The technique searches for "observable implications of hypothesized causal processes" and parallels the way a detective uses clues to solve a crime or a physician uses observation to make a diagnosis (Bennett, 2008). Both the detective and the physician rely on prior knowledge to ground the inference in contextual meaning (Collier, 2011). Prior knowledge is the line that enables the dots to be connected rather than remain as isolated observations.

The social sciences use process tracing for inductive theory generation in qualitative research. The hard sciences use process tracing to explain phenomenon such as the extinction of the dinosaurs or origins of the universe (Bennett, 2008). While the method has actually been debated in academic circles for over a hundred years, it was discussed thousands of years ago by Greek historian Thucydides (Bennett & Checkel, 2012). Although the field of comparative politics receives the recognition for the development of process tracing, it was the field of cognitive psychology that coined the term in the 1970s (Hogarth, 1974). In 1979, a Stanford University political scientist introduced the term for use in developing historical explanations of political systems and then he later specified that process tracing was also useful in making inferences for macro-level explanations (George & Bennett, 1979). The researchers in the field of comparative politics ushered in a recent renaissance in qualitative methods when they proposed process tracing and inductive procedures were rigorous methods; and though these are considered quite different than covariance analysis, they "must be guided by different criteria for analytic rigor" (Steinberg, 2015).

Bennett offers a more detailed definition: "process tracing refers to techniques for examining the intermediate steps in cognitive mental processes to better understand the heuristics though which humans make decisions" (Bennett & Checkel, 2012). Critics have noted that historical narratives generated from process tracing have the potential to be reduced to little more than lazy storytelling (Norkus, 2005). Too often, the approach to process tracing is exceedingly informal, lacking in explicitness of execution, and void of transparency (Mahoney, 2015). In response to valid criticisms and to bolster the systematic process, a set of ten criteria is proposed as necessary to encourage good process tracing techniques that are also transparent (Bennett & Checkel, 2012). By defining and operationalizing a checklist, Bennett has very recently established a starting point to guide process tracing toward quality outcomes:

- **Cast the net widely for alternative explanations.** Process tracing becomes unconvincing when a reader identifies a plausible explanation that the research neglected to address. There could be a bias from an omitted variable or a broader force in play not considered.

- **Be equally tough on alternative explanations.** This refers not to investing the same amount of time per investigation, but to taking the necessary steps to negate confirmation bias. Developing an outline provides the structure to systematically evaluate the evidence for each alternative explanation. For example, when *proximity* was being evaluated in the earliest *intentional communities* during pre-colonial times, that characteristic was accounted for by virtue of the fact Native American raids were a constant threat; any social benefit from that design was secondary as *proximity* was simply a matter of survival.

- **Make a justifiable decision on when to start.** This is a matter of debate but a common strategy is to select a turning point in history. Selecting a long period of time allows for the development of a consistent trajectory and demonstrates a practice or characteristic has been reproduced. For this dissertation, the starting point was when Thomas More published his book *Utopia* in 1516, which created a literary genre that sparked the building of *intentional communities* (Claeys, 2010).

- **Be relentless in gathering diverse and relevant evidence, but make a justifiable decision on when to stop.** Using Bayesian logic and the four empirical tests (hoop tests, smoking gun tests, straw in the wind tests, doubly decisive tests) the probative value of the data not yet obtained can be ascertained. The higher the probative value, the more time it warrants investing. Also, utilizing distinct sources of evidence allows for triangulation, which validates the evidence. Ironically, a vast amount of evidence can serve to reinforce a bias and overlook other crucial evidence. A sensible place to stop is when the repetition of the evidence is obvious and that further investigation will not change the justifiable conclusion.

- **Consider the potential biases of evidentiary sources.** Ignoring this influence allows a naivety to overtake the analysis and false inferences to be made easily. By applying a two-step Bayesian analysis, this bias can be addressed. The first step is to access the possible instrumental motives of people providing evidence, and then to weigh the evidence; the second step involves updating expectations regarding those motives.

- **Take into account whether the case is most or least likely for alternative explanations.** A researcher holds prior expectations on strength of a theory and when the analysis fails to provide an explanation, process tracing can identify if it was a fluke or if the conditions need a radical revision.

- **Combine process tracing with case comparisons when useful and feasible for the research goal.** Comparative case studies can identify omitted variables and evaluate differences that can explain different outcomes. There is useful synergy between case comparisons and process tracing.

- **Be open to inductive insights.** Stumbling across an unobvious insight is a tremendous advantage of process tracing. It requires the researcher to pay attention to surprises in the narratives and then attempt to explain those surprises theoretically.

- **Use deduction to ask, "If my explanation is true, what will be the specific process leading to the outcome?"** This leads the researcher to explore what other characteristics would be evident in a given context if the explanation were true. This would be looking for the proof in the pudding or seeing beyond flattering statements offered as window dressing. If true, there should be corroborating evidence.

- **Remember that conclusive process tracing is good, but not all good process tracing is conclusive.** When a high level of confidence is not possible, the researcher should acknowledge the level of uncertainty. One of the pitfalls to resist in qualitative research is the tendency for curve-fitting to make inductive insights more believable.

Process tracing has traditionally been a more intuitive practice, yet systematic procedures have recently evolved to make it a more rigorous methodology. The first step is establishing correlations. In this dissertation the data were reviewed for specific evidence in the narratives that reflected presence of each theme. The second step is to investigate causality. This involves considering potential spuriousness and considering if X caused Y or Y caused X or if some third variable such as Z caused both X and Y to appear. What is often referred to as "Hume's argument" states the frequent conjoint occurrence of X and Y variables is the essence of causal inference (Bennett & Checkel, 2012). Hume lacks a mechanism though and this gap creates the opening for process tracing. Still, only inferences can be made about causality rather than actually observing causality (Bennett & Checkel, 2012).

Critics can raise the issue of the 'degrees of freedom' problem where the number of potential variables are great but that is justifiably countered by acknowledging that all data are not created equally (Bennett & Checkel, 2012). While three major themes were explored, there are likely more minor themes, but research reflects these three themes investigated are the most prominent and consistent. A researcher who is intimately familiar with the background conditions of certain places and times is in a qualified position to be able to judge the plausibility and significance of each piece of evidence and determine whether more investigation is required (Clayton, 1996). Being part of the sustainability industry developed this useful familiarity. During theory building, other variables were entertained in the spirit of open-mindedness (Mahoney, 2015). There were pragmatic limits to narrow the focus to an answerable research question so as to avoid potential "infinite regress" (Bennett, 2008).

Empirical Tests

Causality in process tracing has a probative value based on four kinds of empirical tests that use a two-by-two matrix to categorize the certainty and uniqueness (Van Evera, 1997). "*Unique predictions* are those accounted for only by one of the theories under consideration, while *certain predictions* are those that unequivocally and inexorably true if an explanation is true" (Bennett, 2008). How the evaluative terms combine creates the four kinds of tests shown in Table 3.

Table 3. Empirical Tests Used in Process Tracing Method

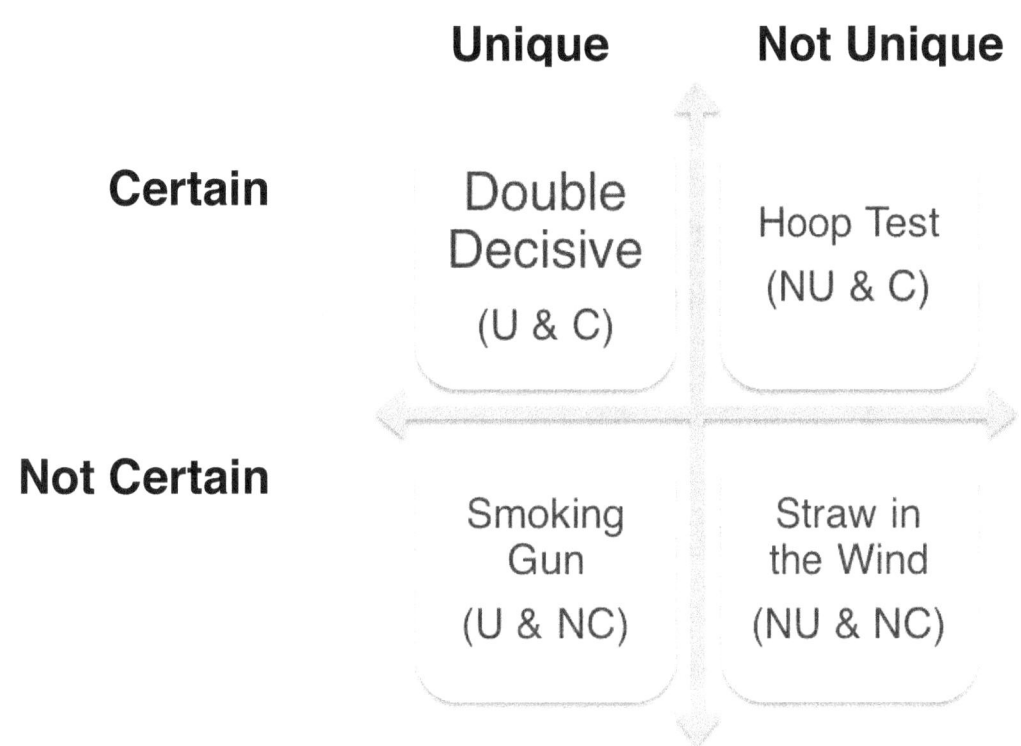

These four categories are not definitive tests but they provide plausibility in the evaluation of causation. The hoop test provides necessary, but insufficient criteria to accept a hypothesis; it can eliminate irrelevant data. Passing the hoop test is required to consider the hypothesis but does not affirm it. The smoking gun test is often sufficient to confirm the hypothesis but absence does not negate the hypothesis. Failing the smoking gun test can have minor or major implications for the hypothesis depending on how difficult the smoking gun test was designed. As the name implies, a gun in a murderer's hand implicates the suspect but absence does not prove innocence. The straw-in-the-wind test does not provide strong evidence for or against the exploration but what matters is the *kind* of evidence it produces and the confidence it allows (Bennett, 2008). Straw-in-the-wind consists of weak and circumstantial evidence yet a series of these tests can increase the confidence in an explanation (Bennett & Checkel, 2014). Lastly, the doubly decisive test is a rare occurrence because it both confirms a hypothesis and eliminates the alternatives. An example of this is a bank camera that photographs the faces of the bank robbers, thus producing convincing evidence and eliminating others of the crime. The combination of a smoking gun and hoop test can create the equivalent of a doubly decisive test (Bennett, 2010). These tests are often used intuitively in the evaluation of evidence or hypothesis plausibility, but the transparent step of articulating them provides a glimpse into the researcher's thought process so the their logic can be followed.

The data were analyzed against four empirical tests before being accepted as evidence for qualitative narrative expression. An attempt was made to conduct a processual analysis of history across hundreds of years of examples of *places of innovation* and *intentional communities*; this massive data collection was done to avoid cherry-picking events to support any preconceived confirmation bias (Bennett, 2008; Mahoney, 2015). Because this research was begun as inquiry-led research, there was actually not a preconceived theory so the evidence gathering process genuinely produced the themes and casted the researcher into the role of a mere spectator of pre-existing patterns. Starting with a clean theoretical slate is ideal if not impossible (Eisenhardt, 1989). Good process tracing creates a transparent logic that discusses how the evidence fits so the reader can reconstruct the logic behind the theory proposed (Mahoney, 2015).

Theory Generation

Process tracing goes through an iterative process that uses Bayesian logic to justify using common intuition (Bennett & Checkel, 2012). Bayesian logic attaches subjective probabilities to the accuracy of the hypothesis and then updates these probabilities as new evidence occurs (Bennett, 2008). Explanations gleaned from process tracing are "provisional" as it is impossible to fully account for all factors that would lead to 100 percent accuracy (Bennett, 2008). Ideally, theory building should begin with no theory in mind because starting with a clean theoretical slate can reduce bias and potential findings (Eisenhardt, 1989).

Once the concepts are generated from the process tracing procedures, the emerging theory is subject to generalization, which is "a type of inference that leverages information and insights to help explain broader collections of social phenomena that they do not measure" (Steinberg, 2015). Because the theory-building process is intimately tied to the evidence, it tends to mirror reality and be empirically valid (Eisenhardt, 1989). There are two kinds of generalizations: within-system generalization and cross-system generalization.

When a hypothesis passes the with-in system generalization, it can then be explored in a cross-system generalization as a future research area in section of chapter 5. The logic of the generalization must show that the findings are believable and could plausibly transfer to a deeper understanding of a different phenomena not studied directly (Steinberg, 2015). Additional theory-testing questions could validate or invalidate the transferability to different settings. A well-specified theory would be useful in offering predictions in different institutional settings over time; because it provides a measure of continuity even as actors and information flows change (Bennett & Checkel, 2014).

The scope of the research for this dissertation was to propose an emerging theory. At the present time, a theory-testing investigation directed toward higher education institutions focused on sustainability-oriented innovations is only hypothetically possible, because no university is currently known to pursue such this specific strategy. Figure 1 shows a graphical representation of how an inductive theory-generating concept is developed for a with-in system generalization and then ultimately applied to a cross-system generalization (Steinberg, 2015). Specific mechanisms were identified in various forms of *intentional communities* and were generalized as an element in *intentional communities*; these are represented by the circles on the graphic. These same specific mechanisms were identified in various forms of *places of innovation* and they, too, were generalized as an element in *places of innovation*; these are represented by the triangles on the graphic. Future research can test an emerging theory originating in a within-system study to a cross-system generalization to determine if what was present in the first analysis is also present in a different setting that combines the observations for a university setting pursing intentional innovation. The potential for future research is discussed in chapter 5 but is illustrated in Figure 1 in the methodology chapter to define

the difference between theory-generating research and theory-testing research.

**Figure 1. Process of Establishing With-in System Generalization
Prior to Applying it to a Cross-System Generalization**

With-in system Generalization
evidence found within the same type

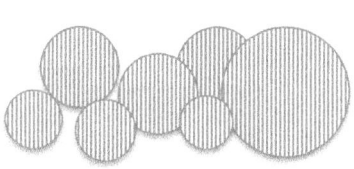 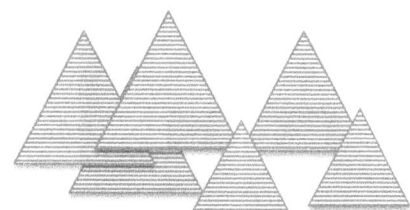

intentional communities:
experimental communities,
communes, ecovillages, etc.

Places of innovation: industrial
districts, clusters, innovation
districts, research parks,
innovation campuses, etc.

Cross-system Generalization
combining generalizations to a new venue

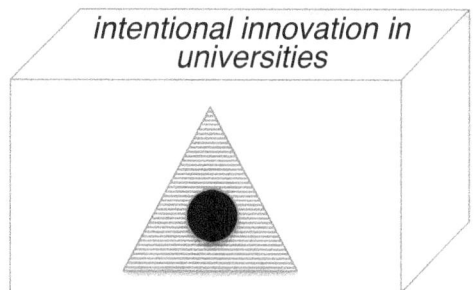

*intentional innovation in
universities*

CHAPTER IV

FINDINGS

This chapter explains the origins of the research questions and provides a succinct

summarization of the findings anchored by references of evidence. The 21 categories of

findings are disclosed in chapter 4 and followed by a discussion of the findings in chapter

5. The purpose of the study was to gather evidence that would help build an emerging

theory about the replicable mechanisms useful to create environments capable of

fostering *sustainability-oriented innovations*. As a supplement to the findings on the

communities of innovators, a section is presented on the demand for *sustainability-

oriented innovation* as found by the sustainability diffusion into three industries: the built

environment, higher education institutions, and business.

Origins of the Research

Initially, my research began a decade prior to this dissertation as a broad investigation

asking sustainability thought leaders what needed to be taught in the university so

graduates could create a sustainable society. This eventually led to research on how to

use university housing as a model of a sustainable community lifestyle and economy.

One of the rare models of a sustainable community in an industrialized country is the

ecovillage. Today's ecovillages ambitiously attempt to create resiliency by addressing

136

housing forms, housing design, community design, transportation, governance, food, energy, water, and the economic base. Understanding the origins of the ecovillage required a deeper investigation into its predecessors, namely the evolution of *intentional communities* established in the United States. The inventive spirit I read about and witnessed in *intentional communities* piqued my interest in venues where innovation was traditionally created by private industry and in venues where universities were involved with innovation.

Surprises

Several findings were surprising in this investigation. First, the big surprise was there was an element of innovative thinking embedded in these communal efforts that was consistently present over time. Prior to doing research on the *places of innovation*, I had focused only on *intentional communities*. Once I followed a rabbit trail into utopian literature and the academic discourse about utopianism, I was able to appreciate their virtues more clearly. Once I understood the historical context from which these experimental communities arose – the possibilities of the New World, the newfound democracy after Revolutionary War, the squalor of the Industrial Revolution cities, the environmental awareness of the 1970s, to name just a few, I saw the nobility and courage inherent within these founders; they were innovators and experimenters.

Communitarian scholars consider *intentional communities* so fundamentally innovative that any discussion of measuring the success or failure of an individual community is discouraged (J. Wagner, 1985). Even short-lived communities are viewed as vital for the iteration needed to support the evolution of this form of settlement (Sargent, 2012). "A completely transcendent utopia is impossible" so any utopian striving is partially successful at contributing to cultural change (J. Lockyer, 2009). The arguments from communitarian thinkers who refuse to label efforts as failures are very similar to the philosophical views of inventors: failures are seen as iterations and, as such, are not deemed failures but considered necessary learning opportunities.

The second surprise was the flipside to the first. There was an element of utopian thinking in the innovation literature. Utopian thinking was the driving force behind the innovations that were created in *intentional communities* and the innovations that were created in *places of innovation*. In fact, I've come to appreciate more clearly that utopian thinking is the manifestation of hope and it is the root of democracy, social justice, equity, and so also the root of many topics in sustainability.

Third, theory-building research itself is an iterative process. It is a dance between ambiguity and confidence. This particular research inquiry was clarified by repositioning it to focus – not on the physical place but rather on the *approach* implemented by communities of innovators regarding their utopian thinking and innovative thinking. The phrase "social nature of innovation dynamics" captures the quest to develop a theory about the underlying determinants of innovation (Bramwell et al., 2012). I also had to accept that even after all the iterations, this dissertation is still going to be D1.0 and the 2.0 version will come in the form of future collaborative (thus a more enjoyable process) publications.

Fourth, this research is full of delightful serendipitous discoveries. There are surprising juxtaposed connections such as those between *intentional communities* and business gurus (such as Stewart Brand) as well as those between communes and *open source* software development (Leadbetter, 2008). A major piece of scholarship about commitment in community came from Rosabeth Moss Kanter's 1968 work on communes (Kanter, 1968). Kanter is now a Harvard professor who has published many books and articles on innovation, corporate change, leadership, and strategy (Kanter, 2000, 2002, 2003, 2006; Moss Kanter, 2012). Paul Hawken is another author who has a connection between *intentional communities* and business publications. His book from 1975 is about his visit to a mysterious settlement in Scotland started by Peter and Eileen Caddy and Dorothy Maclean in 1962. Later in the 1980s, this settlement attracted more residents and became known as Findhorn, earning the distinction of being the first "ecovillage" in the world (Hawken & Herr, 1975). Hawken honed his sustainable business perspectives by launching a wholesale organic food business and later a mail order catalog of garden supplies. Along the way he established himself as a global thought leader in sustainable business by publishing the books *Growing a Business, Natural Capitalism, The Ecology of Commerce, Blessed Unrest: How the largest movement in the world came into being, and why no one saw it coming,* and an entirely updated version, *The Ecology of Commerce* revised edition 2010 (Hawken, 1988, 1993b, 2007, 2010a, 2010b). To me, the fact that these 1960s and 1970s settlements on the fridge attracted so many innovative minds is a testament to the statement these settlements made.

Fifth, I learned the origins for today's global *open source* business philosophy, which grew out of the 1970s *open source* software philosophy, began in the 1960s communes in the United States (Leadbetter, 2008). In the 1960s and 1970s, before commercial software was available, programmers wrote their own code and shared it freely between academic and corporate laboratories. This communal behavior became a central feature of the hacker culture (Hippel & Krogh, 2003). Ironically, many of these programmers had experiences living in communes where they were exposed to a communal ethos (Leadbetter, 2008). The communes inherited, likely unknowingly, their ethic of equality and sharing from the Protestants who protested against the authoritative structure of the Catholic Church centuries earlier (Sargent, 2012). But mass-collaborative Wikipedia should pause before it thanks the Protestants for spawning this communal ideal because equality originates in egalitarianism, which has existed for many centuries. The true origins of utopianism actually come from the earliest Greek and Roman myths written and orally passed down as stories of self-determination, taking control of our dreams, and the "essential need to dream about a better life" (Sargent, 1994).

And lastly, it was the small coincidences that provided me with reprieves of delightful moments. I discovered the required reading list from the University of Wisconsin's 1927 Experimental College contained *The Tragedy of Waste* (Chase, 1925). I had serendipitously discovered this vintage economics book by admiring interesting old books in a forgotten corner of the university library. The title intrigued me; the contents captivated my attention, it contained vast amounts of wisdom. I selected it for the reading list in my own experimental doctoral degree. The book was about the tremendous benefits realized through the efficient logistic management of scarce resources caused by WWI and how tragic it was that same level of responsible use of resources was not pursued during times of peace and abundance. Though the wartime paradigm of frugality decreased pollution and even raised employment, those extreme efficiency measures did not appeal to the policymakers once the war was over. It was an early lesson for me in possibilities and in the power of political will for environmental progress or for ecological folly. Another delightful surprise was around Disney World's EPCOT center. I've visited EPCOT twice, but didn't learn the history of the *Florida Project* and its ties to Garden City. I've queried dozens of thought leaders about the potential application of the Garden City philosophy to no avail. After a decade of searching, I nearly gave up finding anyone as inspired as I was by *Garden City of To-Morrow,* then I met the president of a university who is equally enamored with the book and is using some Garden City philosophies in the their new innovation campus development. This reminds me that great wisdom exists in our history; it is there for anyone to discover.

Research Question

Upon circling back to the original investigation of what needed to be taught, it eventually became apparent that to achieve sustainability, a multitude of *sustainability-oriented innovations* around products, processes, and policies would need to be generated. Subsequently, to understand higher education's past involvement regarding the use of specialized places to generate innovations, led to gathering evidence from *places of innovation:* namely, the research parks and innovation districts. These forms had predecessors dating back to the early twentieth century industrial districts and, more recently, clusters. Not all forms in the *places of innovation* specifically required the involvement of universities. These research investigations focused on identifying the kind of thinking that produced innovation in the past—be it an industry or a new settlement— and how higher education could use the environments they provide, to develop in students the kind of thinking that could lead to innovations for sustainability. Utopian thinking was identified as the basis for the belief in progress and it was traced through the *intentional communities* scholarship but was also inherent in *places of innovation.*

The findings provided data to explore the broad research question: "What are the mechanisms historically used by *communities of innovators*, as identified in *intentional communities* and in *places of innovation*, that were used to approach their goals?" To offer implications from the findings, the research question was then tailored for the university setting: "How can these mechanisms be applied to the environments created by higher education institutions so they can successfully fuel innovations that advance sustainability?"

Three Themes

The key findings led to the recognition of three themes consistently present in the two fields of places known for an experimental nature: *intentional communities* and *places of innovation*. In all, eleven types of *communities of innovators* were considered: *Intentional communities* (also known as experimental communities), communes, ecovillages, academic ecovillages, cohousing, innovation community prototype, industrial districts, clusters, research parks, innovation districts, and universities.

The first theme is referred to in the Computer Age as '*open source* information exchange', a term that captures the collective contributions of user groups voluntarily committed to improving software. This *open source* concept is known for its generosity of spirit and for its benefit for the common good. The second theme is the *iterative process*. This is a design element often woven into the culture of the venture or community. An inherent attribute of the *iterative process* is the deliberate strategy to tolerate and learn from mistakes in the pursuit of a higher ideal. The third theme is the value placed on *proximity*. Based on social networking, shared goals, and physical vicinity, *proximity* forms the working basis for *open source* exchanges. *Proximity* builds relationships of trust necessary to navigate the *iterative process* that allows innovation to emerge from collaborative efforts. The three themes of *open source*, *iterative process*, and *proximity*—when intertwined—are mutually reinforcing.

Establishing units of analysis to measure utopian thinking and innovative thinking is a challenge because of the unobservable nature and causality issues. The unit of analysis is a narrative description that is reflected in the thinking as manifested in the theme. The findings have been derived by identifying these themes in scholarly literature and other published sources of evidence through the process tracing method applied within the historical methodology. Table 4 reflects the quantities involved in the eleven types of *communities of innovators* in the United States, unless otherwise indicated.

Table 4. Quantities of People in Communities of Innovators

Community Form	Date	Quantity Involved
Intentional Communities	now	> 1,000 settlements
(1)Communes	1970	750,000 residents
Cohousing	1988- now	9,500 households
Ecovillages	1973-now	400 villages worldwide
Academic Ecovillages	now	2 villages
Innovation Community Prototypes	1965-now	2 places
Industrial Districts	1900-now	100s of regions in Italy
Clusters	1990-now	50+ of regions
Research Parks	1950-now	195 places
Innovation Districts	2010-now	12 – 80 places
Innovative Universities/Campuses	1930 – now	50+ places

Between 1780 and 1860, there were 90 *intentional communities* founded in the United States and, by 1914, at least another 200 were established (Claeys, 2011). During the 1960s, an estimated 10,000 communes were established that involved 750,000 U.S. citizens (T. Miller, 1992). Households in cohousing communities numbered 9,500 as of 1988 (Williams, 2008). There are about 155 existing cohousing communities in the United States and around 100 more in the planning stages (A. Alexander, 2015). There are over 400 ecovillages worldwide (A. K. Bates, 2003). There have been two academic ecovillages built on university property in the United States (Cal-Poly, 2015; Van der Ryn, 2012). The two innovation community prototypes investigated were Disney's original vision for EPCOT in 1966 and Sim van Der Ryn's Marin Solar Village in 1979. According to the Association of University Research Parks, there are about 150 research parks in the United States (AURP, 2016). There are less than a dozen innovation districts on the ground but there are possibly over 80 cities in the planning stages of an innovation district initiative (Huggett, 2014). In the university category, there are about a dozen universities operating an 'innovative campus' initiative; included in this category are the universities renowned as innovative universities or entrepreneurial universities. Although not listed on Table 4, of the hundreds of experimental colleges begun between 1927 and 1975, there were still 312 viable programs remaining by 1999 (Kliewer, 1999).

Early in the research, the general themes began as subtle 'qualitative revelations'—that then solidified through gradual, but consistent reinforcement as the evidence for them unfolded with each iteration between data and working hypothesis. The connection crystalized with the contemplation that comes with unrushed time and patient analysis. Furthermore, the prevalence of these themes across various forms of communities implicates them as foundational. These themes can frame values, assess progress, and guide higher education institutions as they access their future roles in the *knowledge economy*.

Open Source Definition

The term *open source software* originated over the past few decades in the computer

software industry (Raymond, 1999). *Open source* describes how a source code that would

have been proprietary is, instead, made freely available for public collaboration so that

the code can be improved (Steele, 2012). Source code, also referred to as coding, is "a

sequence of instructions to be executed by a computer to accomplish the program's

purpose" (Hippel & Krogh, 2003). The intent of making a source code open is to take the

original source code, modify it, and then share the improvements back with the

community of software developers (Perens, 1999).

Open source (or open-source as it is sometimes written) is a descriptive term that has

now been extended to apply to any initiative that promotes the values of open exchange,

collaborative participation, transparency, truth, trust, and even rapid prototyping (Steele,

2012). These qualities are now considered proven catalysts for innovation (DiBona &

Ockman, 1999; Goldman & Gabriel, 2005). A recent evolution of *open source* is the

paradigm known as *open innovation* (H. Chesbrough, 2006; Chesbrough & Bogers,

2014). But long before the Computer Age, the value of openly sharing information,

knowledge, and wisdom was integral to other start-up efforts such as the communes of

the 1960s (Leadbetter, 2008). In the early twentieth century, economist Alfred Marshall

described industrial districts as possessing a ubiquitous nature of expertise that saturated

a region as being 'in the air' because it permeated the entire culture (Belussi & Caldari,

2009).

The sharing motive behind the *open source* approach has its roots in a wider movement referred to as *commoning*: "to produce or extract a livelihood from a common resource" (Linebaugh, 2008). Information flows freely between stakeholders because there is recognition that the collective genius of a larger group can create a better product than an individual or a smaller group. These collaborative initiatives are evident in the crowdsourcing approach to raising funds (Doan, Ramakrishnan, & Halevy, 2011). In 2009, economics professor Elinor Ostrom was awarded a Nobel Prize in Economic Sciences for research showing that the commons could effectively be managed through cooperation and self-governance (Hardin, 1968; Ostrom, Walker, & Gardner, 1992). Previously, for half a century, it was believed any property held in common ownership and shared would succumb to overuse and depletion as professed in Garrett Hardin's famous essay *The Tragedy of the Commons* (Hardin, 1968). In contract to Hardin, Ostrom used research gleaned from working with actual communities to develop eight principles that guide how resources could be shared for mutual benefit (Ostrom, Burger, Field, Norgaard, & Policansky, 1999). *Open source* software programming, cohousing, collaborative fundraising, and commonly-managed resources—while on the edges of contemporary societal practices—still reflect an ethic of sharing (Pickerill, 2015).

Iterative Process Definition

The *iterative process* originates in the field of engineering (Archer, 1964). Figure 2 reflects the engineering design process. There are discreet steps that, when pursued in order, create a feedback loop of continuous improvement. Today, this problem-solving technique is used in multitude of fields and has become a foundational teaching pedagogy in innovation programs and design schools such as the d.school at Stanford University.

Figure 2. Iterative Engineering Design Process

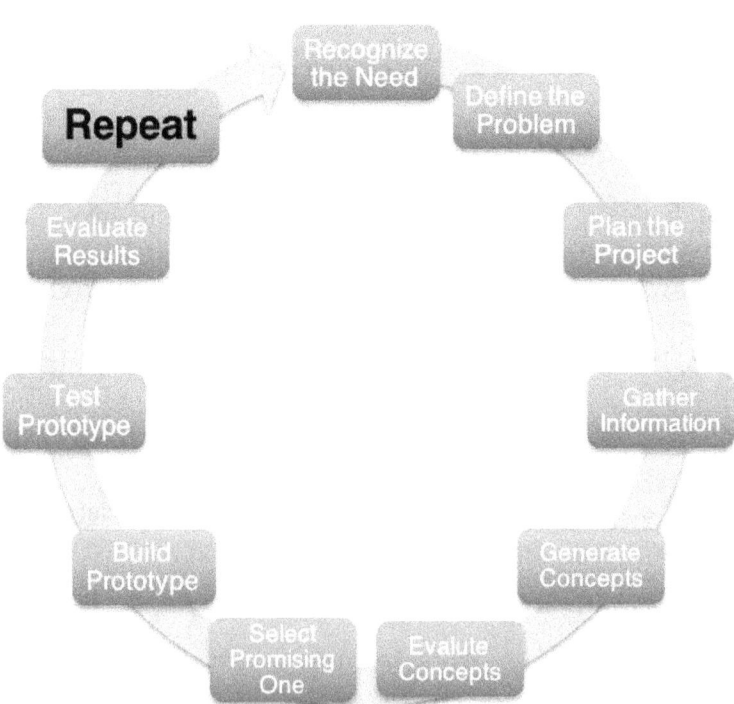

Proximity Definition

For the purposes of this research, *proximity* in *intentional communities* and *places of innovation* is defined as *physical proximity*, *functional proximity*, and *emotional proximity*. Each type of proximity provides a specific function. Physical proximity is a geographical measure of distance or travel time and this allows for those serendipitous chance encounters where tacit knowledge can be shared or 'spillover'. Functional proximity addresses the intentional design of a place such as common areas designed into cohousing or a public realm found in an innovation district. Emotional proximity—the most intangible, yet vital—captures the spirit of the culture and the commitment to community ethos. Without emotional proximity, the other two types of proximity cannot contribute to collaboration; there is no engagement, no trust, and no progress.

Tacit knowledge is unique to place; it is a function of the mix of people, resources, culture, and goals. Serendipitous moments happen at the intersection of *proximity* and tacit knowledge. Often scholars in the innovation research field emphasize how important face-to-face communications are for building relationships of trust that serve to foster the development of an innovative culture in a specific place. This is why the corporate co-location of industry on campus and intermingled with university faculty, staff, and students is a prominent feature in the strategy of innovative universities. In a world where competition from online courses is fierce, there are certain attributes of innovation that can only happen 'in place'; thus flipping the advantage to the brick and mortar universities.

Categories of Evidence

Table 5 summarizes the categories o findings that reflect how the three themes manifested in both *intentional communities* and *places of innovation*. Listed under each type of community are the themes *open source*, *iterative process*, and *proximity* and the type of evidence that supports that theme. All together, there are twenty-one categories of evidence explained in the analysis that follows.

Table 5. Evidence for the Three Themes in Communities of Innovators

Themes in *Intentional Communities*
• Open source: Communal property
• Open source: Governance based on consensus
• Open source: Social glue in cohousing, ecovillages, communes
• Open source: Social glue to generate innovation in EPCOT
• Iteration: Transformation of the individual resident
• Iteration: Community adaptation to external culture
• Iteration: New forms of intentional community
• Iteration: Utopianism, an iterative nature
• Iteration: Governance by consensus building
• Proximity: Physical, land use leveraging vicinity
• Proximity: Functional, bioregionalism for resource efficiency
• Proximity: Emotional, aligned values

Themes in *Places of Innovation*
• Open source: Industrial districts as product of the local culture
• Open source: Innovation districts embedded in ecosystem & community
• Open source: University experimental, spillovers, and corporate co-location
• Iteration: Industrial districts with a culture of constant improvement
• Iteration: Research parks transforming into mixed-use communities
• Iteration: Innovation districts planning and building over time
• Proximity: Physical - benefits of density, spillovers, and serendipity
• Proximity: Functional - access to tacit knowledge
• Proximity: Emotional - builds trust

Communities of Innovators in Intentional Communities

Utopian thinking was a fundamental drive in *intentional communities* and there were mechanisms or approaches that fostered this kind of idealistic thinking. The goal of this dissertation is to identify the underlying determinants that fostered the evolution of utopian thinking in these various types of *communities of innovators* found in *intentional communities* and in *places of innovation*. The evidence of these determinants was called themes. The three themes of *open source*, *iterative process*, and *proximity* were present in narratives found in intentional communities, communes, cohousing, ecovillages, academic ecovillages, and innovation community prototype.

Open Source Found in Intentional Communities

Communal property.

The *open source* theme is prevalent in *intentional communities* as evidenced by the ownership structure providing for communal property. Early intentional community settlements had an ownership structure that was based on commonwealth principles, which basically implies a voluntary association with an enterprise that is mutually beneficial to all. Over the centuries, as town development costs increased with paid land acquisition, the ownership structure became a mixed model of public and sole proprietorship (Carmony & Elliott, 1980). In the case of New Harmony, Indiana, an intentional community in the early nineteenth century, this lack of communal investment became a point of contention among the members (Carmony & Elliott, 1980).

151

They felt their labor contributions were not reflected appropriately with financial recognition; this became just one of the reasons the community ceased after three years (Carmony & Elliott, 1980). As for communes, communal property ownership was empirically determined to be one of the keys to their long-term survival and success (Bader, Mencken, & Parker, 2006). *Intentional communities* have long been recognized as having a collectivist nature (Levitas, 2010; K. Mannheim, 1936). Even the modern cohousing models, which are structured for private ownership of homes, almost always require the common house and public realm to be held in joint tenancy (Jarvis & Bonnett, 2013). Eco-communities and ecovillages reflect sharing, interaction, and mutual support in an attempt to "materialize the commons" (Pickerill, 2015). Valuing the commons, also known as *commoning*, is a practice used in many situations ranging from *open source* software development to village design to Wikipedia contributions.

Communal ownership of real property mimics an operational approach in much the same way *open source* software development does; it reflects an investment in the product which, in this case, is the land, buildings, and revenue streams. With the investment comes the inclination to protect it and increase its value by contributing further improvements. Because the real property is owned communally, there is no proprietary barrier and each owner has permission to contribute to improving the investment. Ownership provides the legitimate right to collaboratively participate in the betterment of the property. With communal ownership—as with *open source*—there is no 'mine' and 'yours', the product is 'ours'. The responsibility then falls to each owner to make the product better for the benefit of the greater society as well as for the contributor.

Governance.

The *open source* theme is also woven into the community governance system of modern *intentional communities* that utilize consensus-based decision-making. The early *intentional communities* of the eighteenth and nineteenth centuries used either a paternalistic decision-making style or appointed a council for governance. More often than not, these early *intentional communities* did not survive and prosper under autocratic governance in a country that was settling into a new democracy (A. E. Bestor, 1950; A. E. Bestor, Jr., 1948, 1953). Through experimentation, this evolved into a new model of consensual decision making (and sometimes a blend of consensus and majority rule) that is currently used in ecovillages and *intentional communities* today (Cummings, 2003). This novel governance structure was incorporated into the fundamental design of the cohousing model from Scandinavia that was introduced to the United States by architect Charles Durrett in the 1980s (McCamant & Durrett, 1994).

The inherent transparency of *open source* is the basis for a discussion-based consensus process. The product is the governance. The dynamic that makes consensus related to *open source* is the recognition that the community, as a collective, is able to cooperate and collaborate to produce a better decisions. *Open source* allows a venue for each member's experience to be heard and that member's intellectual contribution can modify and improve the decision under consideration by the group. Consensus decision-making is a collaborative exercise of open participation based on trusting other members enough to share a personal opinion. Many of the modern *intentional communities* train their members in this egalitarian decision-making process prior to constructing the actual community (Durrett, 2015). From contributing equally in the initial design phase of the community layout and architecture to the day-to-day governance, it is this *open source* approach that allows these *intentional communities* to be co-created by the community of residents in much the same fashion that open source software is co-created by the community of developers and users.

Social glue.

Open source is also seen as the intangible social glue of relationships that evolve over time and are maintained through healthy communications. Cohousing researcher Jo Williams explains:

> Local social capital is the 'glue', which binds people together in a neighourhood and encourages them to cooperate with each other. It is the local networks together with shared norms, values, and understandings that facilitate cooperation within or among groups in a neighborhood. Without social capital individuals feel isolated and are untrusting, which reduces levels of cooperation within a neighborhood. (Williams, 2005)

The ability to be open and trusting with neighbors is what makes a collection of homes into a community of tight-knit neighbors (Andreas & Wagner, 2012). It is common for intentional community members to be trained in non-violent conflict resolution to facilitate open productive communication (Schehr, 1997). In the absence of guarded exchanges and personal secrets, the community can avoid creating a lack of information. They can, instead, choose openness and trust to create familiarity, which engenders compassion for each other and builds the solidarity of the community. Meaningful communication and personal relationships are the lifeblood of *intentional communities* (K. T. Litfin, 2014). *Open source* is the platform for transparency that allows for a generous sharing of intellectual capacity and personal resources, thus facilitating crucial communication for the continuous betterment of the community. The product being collectively created by the community is the community spirit or the malleable social glue.

This social glue that operates internally is also echoed in how the intentional community engages in the context of the larger regional surroundings. Modern ecovillages often fill the role of educational outreach to the surrounding region. They position themselves as model demonstration facilities (Allen-Gil et al., 2005; Andreas & Wagner, 2012; Walker, 2012b). The *open source* approach is reflected in the porous nature of how information flows from and back to the community; the intentional community is continuously learning from those on the outside, experimenting with those technologies internally, and sharing what they have learned in practice back to the regional community (F. Wagner, Andreas, & Mende, 2012). Rather than 'hiding their source code' for sustainability, the twenty-first century ecovillage openly shares their lessons learned and best practices for the overall betterment of other ecovillages as well as individuals pursuing a more sustainable lifestyle (Tinsley & George, 2006).

At the organizational scale, this social glue found in the *open source* approach is present in the trade organization that facilitates the networking of the *intentional communities*. A collection of intentional community advocates first loosely banded together in the 1940s. The current global organization, The Fellowship of Intentional Communities, was founded as a non-profit in 1986; its primary purpose is to facilitate the collective evolution of this type of settlement (FIC, 2016). They publish a monthly magazine, maintain a list of *intentional communities*, and hosts global gatherings for the purpose of networking the various *intentional communities* so they can share best practices.

Social glue in EPCOT.

Project Florida, as designed, was never built. But based on the few public statements and what is known about *Garden City of To-Morrow's* Ebenezer Howard's influence on Walt Disney, using an *open source* approach would have fit with Disney's intention of creating a place that fostered a feedback loop between the innovation and the users. Disney was famous for putting families first in every design and project he created (Chytry, 2012). Above all other goals, Disney's purpose was the happiness of the community (S. Mannheim, 2012). Disney's design for the Experimental Prototype Community of Tomorrow *intentionally* mixed citizens with industry inventors (Nachman, 2014). He even designed a monorail transportation system to circulate people easily and, presumably, circulate their ideas.

Disney operated under the intuition that the ultimate *knowledge spillover* would occur in a perfected public realm between people who lived with the technology and those who invented it. In the *Project Florida* film, Walt Disney explained to his premise:

> E.P.C.O.T will take its cue from the new ideas and new technologies that are now emerging from the creative centers of American industry. It will be a community of tomorrow that will never be completed, but will always be introducing, and testing, and demonstrating new materials and new systems. And E.P.C.O.T will always be a showcase to the world of the ingenuity and imagination of American free enterprise.
>
> I don't believe there is a challenge anywhere in the world that's more important to people everywhere than finding solutions to the problems of our cities.
>
> But where do we begin? How do we start answering this great challenge? Well, we're convinced we must start with the public need. And the need is not just for curing the old ills of old cities. We think the need is for starting from scratch on virgin land and building a special kind of new community. So that's what E.P.C.O.T is: an Experimental Prototype Community that will always be in the state of *becoming*. It will never cease to be a living blueprint of the future where people actually live a life they can't find anyplace else in the world.
>
> There will be no retirees; everyone must be employed. (S. Mannheim, 2012).

Evidently, Disney wanted every citizen contributing by being employed. He stressed the community would be in a perpetual state of '*becoming*'. The intention of the community was to be "introducing, and testing, and demonstrating new materials and new systems." The community component was intended to act as a beta city or a testing ground. Disney often alluded to the plethora of ideas he carried in his head and one has to wonder if this culmination of his life's work was an effort to create a place for people overflowing with ideas—like himself—to bring those ideas to life and test them.

Iterative Process Found in Intentional Communities

The theme of *iterative process* is woven into *intentional communities* due to their alignment with utopian aspirations. The residents continuously evolve, each community adapts, and new forms in the built environment constantly emerge as the movement responses to the culture. The evidence of *iterative process* is presented in five categories: individually, community adaptation, new forms, utopianism, and governance.

Individual transformation.

The *iterative process* is a theme found in *intentional communities* concerning the development of the individual. There is an undeniable individual transformation that takes place at the member level inside an intentional community (S. L. Brown, 2002). Sustainability within an ecovillage is a personally transformative process that makes global sustainability more imaginable (Hong & Vicdan, 2015). This self-transformation is intentional and one of the main draws of participating in an intentional community (Schehr, 1997). *Intentional communities* recognize they are not just building a new style of neighborhood or settlement but they are, in fact, "building new people" (McLaughlin & Davidson, 1985). Alumni of the experimental communities in The Ecovillage at Ithaca and Arcosanti in Arizona both report that as newcomers, they found they had to reinvent themselves to fit into the radically different social structures of established ecological communities (Bochinski, 2016). The ability of an individual resident to adapt to the evolving community persona was identified as a key success factor in the longevity of a community (Kanter, 1972). Between the tasks of redesigning many facets of society and transforming themselves, the intentional community members are in a constant state of iteration as they re-imagine themselves, re-imagine their community, and envision a better society at large (Hong & Vicdan, 2015).

Within the ecovillage domain, the stated purpose of an ecovillage is to develop a lifestyle with a consciousness of an ecological budget. The implied purpose is to provide a venue for an individual to continuously adjust their lifestyle to align themselves closer to sustainability ideals; this is described as an on-going effort and difficult process, but a worthy pursuit (Andreas & Wagner, 2012; K. Litfin, 2007; Mychajluk, 2014). Sustainability is often described as a journey, not a destination, and along that journey there are many opportunities for self-examination and iteration through choices that gradually reduce the ecological impact of an industrialized lifestyle.

Community adaptation.

The *iterative process* is found in *intentional communities* with regard to the adaptation to the external culture. According to Jared Diamond, author of *Collapse: How Societies Choose to Fail or Succeed*, a given society's ability and willingness to adapt to its external surroundings has been empirically established as a valid measure of survivability (Diamond, 2005). Most of the modern ecovillages have embraced their lifecycle; they evolve from struggling start-up communities, establish themselves, and then morph into established communities with outreach venues. The settlement as a whole recognizes the need to adapt to the ever-changing goals of their current members. Historically, settlements that refuse to adapt hasten their demise and eventual collapse. Adaptation of the settlement to the external society is the final step of a continuous effort in the *iterative process*. Successful adaptation signifies feedback was incorporated and changes were implemented. The practices of the past were adapted to change into a more effective—or sustainable—form. Entire civilizations that could not change, would not change, or refused to voluntarily adapt have collapsed. The wiser *intentional communities* accept that the only constant is change and they navigate their community through adaption to the external culture.

Scholarship within *intentional communities* developed theories that explain *intentional communities* are purposefully structured to accommodate an adaptive process. One such theory was Donald Pizter's *developmental communalism* theory that generated seminal insights into the process of change within the community as a whole (Pitzer, 1989). As an intentional community continues toward its goals, it allows alterations by permitting some of the original goals to dissolve so the community can fit into the broader social movements occurring in society (Pitzer, 1989). This adaptation feature is unique to the intentional community whose purpose from the outset goes beyond mere communal housing. The intentional community sets idealistic goals and then iterates how those goals are prioritized and pursued in response to new challenges and opportunities (J. Lockyer, 2009). As members iterate better versions of themselves, their communities also evolve as a by-product of those personal transformations and the communities themselves iterate into new forms of "infinite improvement" (Schehr, 1997).

In the case of high-turnover in *intentional communities*, the residents have learned to successfully adapt to the reality of 'stayers and seekers' and this perpetual influx and outflow is a prompt to continuously iterate the community identity based on ever-changing contributions (Aguilar, 2012). Though few communities plan for instability at the outset, a balance can be achieved between high-turnover and membership stability "by recognizing the increasingly bifurcated membership and adopting behaviors and ideologies that address the needs of both groups" (Aguilar, 2012).

Building on Pizter's *developmental communalism* theory, Lockyer interpreted it with more explicit terms and expanded it into his own *transformational utopianism* theory because he proposed that the term 'transformational' more accurately described what he had witnessed in modern ecovillages. Lockyer describes the iteration as "a process of imagining and attempting to enact more ideal social forms within both individual *intentional communities* and broader social movements" (J. Lockyer, 2009). Lockyer asserts that the experience exceeds the confines of conventional development and is, in fact, more aligned with an intense transformational experience of the overall community.

People are drawn to *intentional communities* to experience personal change, the individual communities transform over time, and history shows that the *intentional communities* movement, as a whole, evolves over eras. Whereas *intentional communities* once withdrew themselves from society to live according to their own private preferences (and still do occasionally), by the 1800s they had evolved into eagerly positioning themselves as models for the entire young nation to replicate (A. E. Bestor, Jr., 1953). The United States was a Garden of Eden, ripe for new people with ideas to develop a new culture. Over time, *intentional communities* evolved to combine their mission and their

161

tendency to retreat, so they became both a safe haven for their members, but also an example for others (Fogarty, 1980). Fogarty's point is that ecovillages only retreat from the world so that they can lead the world by example.

Specific examples of iteration at the community scale are found in the ecovillage case studies about The Farm in Tennessee and Earthhaven in North Carolina (J. Lockyer, 2009). The Farm was founded in 1971 and soon determined their original vision was unobtainable due to lofty goals and limited resources. No longer based on communal income sharing, the private income structure allowed The Farm to evolve into the functioning successful entity that it is today (J. Lockyer, 2009). Earthhaven, an ecovillage established on 325 acres of land in 1994, was inspired by the nearby Celo Community and the learning process Celo experienced inherent in this type of utopianism (J. Lockyer, 2009). *Intentional communities* typically begin with very high ideals, but reality tempers their utopian idealism. Ultimately, they create more transcendent models, using lessons and models provided by previous *intentional communities* (J. Lockyer, 2009).

Social movements are society iterating itself to address a deficit in the culture. Ecovillages are considered social movements because they employ collaborative and communal orientations (Schehr, 1997). These orientations are a radical departure from the "treadmill of production" that requires increasing material possessions that overtax the limited natural resources to support a constant growth economy (Buttel et al., 2004). This social critique is manifest in a lifestyle that is a perpetual, slow iteration and "persistent negotiation" between the dominant culture ideology and the sustainability goals of the community (Ergas, 2010). Systemic culture change overnight is virtually impossible, but the ecovillage purposefully iterates how it operates in the world and tries to balance that tension.

New forms evolve.

The *iterative process* has propelled *intentional communities* to continuously evolve into new forms in the built environment. The pre-colonial *intentional communities* were based in venues based in entrepreneurial ventures or venues providing freedom of religion. After the American Revolution, the experimental community movement manifested in hundreds of settlements involving over 100,000 citizens (Claeys, 2011). While conventional *intentional communities* continued to form into the twentieth century, new forms also emerged in response to the dynamic cultural changes. After WWII, student co-ops emerged, then the hippy communes lured 100,000 people during the social unrest of the 1960s, soon the ecovillages surfaced in response to new environmental awareness in the 1970s, and eventually cohousing materialized in the 1980s. While many universities engage in sustainability-themed dormitories, only a couple of universities have ventured into providing a new type of housing along the lines of an 'academic ecovillage'. The evolution of form has led to a wide diversity of what constitutes an intentional community; the *iterative process* is a constant, particularly in this type of settlement that exists as a social critique of mainstream society.

163

Oscar Wilde said, "Progress is the realization of utopias" (Wilde, 1950). Progress means

today's version is more desirable than yesterday's version, thus progress through utopian

accomplishments is the result of iteration, not accident. Intentional community scholar

Lyman Tower Sargent regards the utopianism expressed through *intentional communities*

to be a "continuing resurrection, reconstitution, and renewal" (Sargent, 1994).

Resurrection implies a cycle of dying and being reborn, all the while learning from

previous intentional community models; this is the *iterative process*.

Utopianism.

Transformation is not an easy linear process; it is a circular iteration of leaving the

original state and becoming something altogether new. As such, the belief system of

utopianism, at its core, is iterative and "should be understood for its enduring and

renewable transformative potential" (J. Lockyer, 2009). The communitarian theorists,

who established the nineteenth century experimental *intentional communities*, were "heirs

of the Enlightenment" who believed fervently in the idea of human progress and

perfectibility (Schafer, 1978). These heirs in 1800 were just 25 years past the

Revolutionary War of 1776. They were still very much living in the same "mainstream of

American thought, which produced the Declaration Independence, the Constitution, and

the Bill of Rights" (Schafer, 1978). The nineteenth century United States citizen

shouldered a huge responsibility; they "expressed the most basic of American ideas and

their concepts and works were carefully studied as portents of America's destiny"

(Schafer, 1978).

These utopian qualities—the idea of progress and the quest of perfectibility—are, quite literally, what drive the *iterative process*. These qualities are now so ubiquitous that they seem obvious in today's culture, but being able to embrace the *iterative process* is fundamental to having a nation known for innovation. Though sometimes a point of debate, "intentional societies are an aspect of utopianism" (Schafer, 1978). Intentional community founders are the consummate inventors of a society that believes the experience of life could be designed better through the trial and error of urban form and governance. In a seminal review of the nineteenth century *intentional communities*, one historian referred to them as "patent-office models of the good society" (A. E. Bestor, Jr., 1953). The explanation that "utopianism is more productively understood as a *process* or method rather than a finished product or an end" is a validation of the *iterative process* inherent in utopian efforts (J. Lockyer, 2009).

The motivation for improvement starts with dissatisfaction of the status quo. The intentional community is an experiment of what could be and is prompted by a critique of the larger society (Abrams & McCulloch, 1976). Utopians view the world as lacking in many ways and their aim is to improve it (Sargisson, 2002). *Intentional communities* are a place for experiments on how to change lifestyles and make dreams come true (Sargisson, 2004). They are not a place of "intellect idealizing" but rather a place committed to praxis (Abrams & McCulloch, 1976). "Process is the hallmark of the non-authoritarian utopia. As such, process—change that does not stop at some ultimate end or goal—requires at the very least a retheorization of intention away from the ends and means" (Garforth, 2009). When change does not stop, the *iterative process* is in constant motion. Accordingly, "Utopia is endlessly dynamic and rampantly generative of the new and the potential" (Garforth, 2009). Clearly, utopians intuitively embrace the iterative nature of development and evolution.

Within the centuries-old intentional community category, the relatively recent emergence of the ecovillage directly reflects the utopian DNA. The ecovillage positions itself as a model sustainable community, designed to demonstrate an alternative lifestyle (Ergas, 2010). Renowned village designer, Christopher Mare, captured the iterative nature of the ecovillage when he says, "the fundamentals are that a sustainable village cannot be created – it must be designed to create itself. The challenge is to design a living system that can assume a life of its own" (Jackson & Mead, 1998). The concept of instilling the implicit design feature of iteration into the explicit intentional capacity of self-generation is an ambitious evolution to place on any village (eco or conventional) but *intentional communities* do have a long-standing commitment to reimagining.

Walt Disney built the Disneyland theme park in southern California in 1955. *Project Florida* was more than an iteration of the California theme park. It was the accumulation of Disney's life's passions mixed with urban planning. Disney, filled with sincere grandeur, explains his vision:

> But if we can bring together the technical know-how of American industry and the creative imagination of the Disney organization, I'm confident we can create—right here in Disney World—a showcase to the world of the American free enterprise system. I believe we can build a community that more people will talk about and come to look at than any other area in the world. And with your cooperation, I'm sure that the Experimental Prototype Community of Tomorrow can influence the future of city living for generations to come. It's an exciting challenge; a once-in-a-lifetime opportunity for everyone who participates. So that's what E.P.C.O.T is: an Experimental Prototype Community that will always be in the state of *becoming*. (Disney, 1966)

'Becoming' implies iteration at the community scale. For all the glory and splendor of the *Project Florida* plans, it was driven by the ancient utopian aspiration of creating a better place that was designed to iterate (S. Mannheim, 2012).

Governance.

Iterative process is a characteristic of the consensus governance model used by *intentional communities*. The *intentional communities* from the twentieth century to present are best known for social innovations in cohabitation, governance structures, and non-violent communication (Schehr, 1997). Within these governance structures, is a variety of decision-making processes with consensus-based being the most widespread. Consensus governance contributes to the attachment cohousing members experience (Brindley, 2003). This attachment creates a stable and engaged membership that practices a consistent commitment to govern their community through iterations of dialogue to reach a decision.

In the United States it usually takes just a majority vote in an election to determine a given course of action, so the consensus-based decision-making process is very novel and unfamiliar to most citizens. Consensus-based decision-making involves voting frequently as a way of checking in with the membership and assessing movement on a topic. Objections are then noted and they become platforms for further discussion (Zablocki, 1971). This *iterative process* compares to rapid prototyping. It thrives on soft skills of transparency, communication, and respect. In consensus-based decision-making, the final decision will not be made until there is 100% consensus (or sometimes a 2/3rd) among the members (Dressler, 2006). The final course of action becomes one that each member can live with while going on record they are 'standing aside'—it is a commitment to proceed in the best interest of the group (Dressler, 2006). Even if some individuals cannot agree with the group decision, they publicly announce they *can* commit to support it for the greater good of the community. Commitment to the *iterative process* is a willingness to listen to dissenting views, consider information from all members, and respect the differences of opinions. The *iterative process* operates on the premise that the collective wisdom of the group is superior to any one individual's wisdom. The consensus-based decision process can go through a few iterations or many iterations and can happen during a single meeting or a series of meetings that last for months (F. Wagner et al., 2012). In 1825, New Harmony used an elected council to make decisions rather than a consensus model, but they had a mandate to iterate governance issues. The process to expel a member was required to go through two public discussion cycles before a

decision was reached (Carmony & Elliott, 1980).

Proximity Found in Intentional Communities

The theme of *proximity* is fundamental to *intentional communities* in terms of physical *proximity*, functional *proximity*, and emotional *proximity*. *Physical proximity* is demonstrated through land use and common facilities. *Functional proximity* refers to strategic bioregional placement. *Emotional proximity* denotes that members' personal values are aligned. The land use design accentuates the physical *proximity* of the housing stock, which fosters the sharing of common facilities; both of these elements—close housing and shared facilities—are purposely designed into *intentional communities* for the express purpose of maximizing social interactions. Bioregional placement is a *functional proximity* because it promotes resource efficiency, which is a priority ecovillages share. And lastly, aligned values refer to the shared goals and collective vision a community holds denoted as *emotional proximity*.

Physical proximity.

In the United States, the earliest European settlements four hundred years ago were designed with physical *proximity* as a priority for survival purposes. Buildings were situated inside a defensive fortress protected by a tall wooden wall on the perimeter. As the threat from attacks from local Native American tribes waned, *intentional communities* continued to foster *proximity* for the sake of efficiency. Clearing forest land was labor intensive and land was prioritized for food production rather than space between dwellings (Carmony & Elliott, 1980). These early agricultural communities featured clustered homes and expansive fields, not yards. Over time, this compact village design came be to be appreciated for its integration of social and work life in the urban setting as documented by centuries of evidence from around the world and across cultures and eras provided in *The Pattern Language* (C. Alexander et al., 1977).

There has been a long held belief in the subtle, but persuasive, power of physical design to impact the social life of a community. The design principle in architecture calls this *physical proximity* approach *social contact design* because it is intended to increase the frequency of social interactions as well as the quality of those interactions (Talen, 1999). Frequency builds bonds; "social interactions within the neighbourhood help to encourage the growth of social capital" (Williams, 2005). Cohousing scholars note that *proximity* fulfills an "enormous yearning to be connected emotionally" (Jarvis & Bonnett, 2013). Though cohousing is traced back to the Swedish collective housing of the 1960s, it "captures the enduring ideals of a much longer communal imagination" (Jarvis & Bonnett, 2013). The need to be located in close *physical proximity* to each other is both primal and preferable.

There are many factors in *proximity* that affect the success of a cohousing community: the number of passive contacts, the density of dwellings per acre, the distance between dwellings, the similarity or homogeneity of the residents, the use of buffer zones, the communal spaces, the overall number of residents, clustered parking, and many other nuanced characteristics unique to each community (Williams, 2005). Like intentional community designers and residents, New Urbanists share "an enduring faith in the power of physical design to change the social life of a community" (Jarvis & Bonnett, 2013). Javis posits that "nostalgia resides at the heart of belonging and attachment" and that the informal everyday practice of sharing meals, sharing ownership and engaging in consensus governance creates nostalgia in cohousing (Jarvis & Bonnett, 2013). Creating nostalgia addresses the need for community, safety and belonging for some idealized community from the past (Jarvis & Bonnett, 2013).

Functional proximity.

In ecovillage design, *proximity* at the regional scale is a practical necessity to achieve self-sufficiency with regard to decreasing the environmental impact of resource acquisition, reducing global dependence, and developing a local economy (Trainer, 1998). This proposed model of bioregionalism expands the ecovillage philosophies of living with less and living more efficiently; it asks the residents to attempt those lighter lifestyles in a local and regional context. Not only are the dwellings in an ecovillage physically close, but the settlement itself is strategically placed near vital resources of water, food, economic bases, transportation and energy capacity (Walker, 2005). By no small coincidence, this recognition to strategically place a settlement in *proximity* to resources harkens back to the sixteenth century. *The Law of the Indies* planning principles were utilized by the Spaniards in their quest to establish towns in the New World. Such *functional proximity* is a timeless concept (Mundigo & Crouch, 1977).

Emotional proximity.

Physical proximity alone is a start, but it is not enough to create strong social bonds; *emotional proximity* is also necessary. *Intentional communities* are formed around specific intentions, which include aligned values, similar mindsets, and common goals.

Trust in relationships is built through frequent, purposeful interactions and is the basis for creating social networks (McCamant & Durrett, 1994). In cohousing, "shared meals, collective ownership of amenities and consensus government" is credited with creating an "attachment to informal everyday practices" or *social architecture* (Brindley, 2003). The social architecture of an intentional community fosters open mindsets so that residents closely align with the community identity (Meltzer, 2005). When members maintain mindsets that differ greatly and are not proximate to each other, it can lead to the collapse of the intentional community, as was the case in the demise of New Harmony, Indiana (Carmony & Elliott, 1980). The commune literature showed that an intangible community spirit can dictate the longevity of the settlement (Kanter, 1968). New Harmony, Indiana demonstrates that the workers and the intellects were not able to reconcile the logistics of the resource distribution or the inequity of the field work, thus the entrenched mindsets led to apathy, resentment, and disbandment of the community (Carmony & Elliott, 1980).

Communities of Innovators in Places of Innovation

Innovative thinking was found in mechanisms used by *communities of innovators* in industrial districts, clusters, research parks, innovation districts, and universities. Each of these types of places hosts innovation dynamics, which—innovation being social in nature—means the evidence must be teased out of narratives describing these places. The other sources of evidence were culled from academic research in economic geography, innovation, and knowledge systems as well as from the promotional publications from innovation districts and universities. The goal was to identify the underlying determinants that lead to the creation and spread of new knowledge that drives innovation that, in turn, is thought to drive economic performance. The underlying mechanisms that foster innovation in *places of innovation* are: *open source*, *iterative process*, and *proximity*.

Open Source Found in Places of Innovation

Open source is a modern term that has evolved over the last two decades; therefore, the mechanism being investigated was not expressed using that exact semantic in the literature. Even so, *open source* was identified as an inherent characteristic in the original industrial district concept, in today's innovation district developments, and in universities contributing to knowledge systems. For reasons discussed in chapter 5, clusters and research parks historically did not utilize the *open source* approach in dealing with their external environment. Before presenting evidence from industrial districts, innovation districts, and universities, an overview of *open innovation* and *open source* provides context for the findings.

Open innovation is defined as "the use of purposive inflows and outflows of knowledge to accelerate internal innovation, and expand the markets for external use of innovation, respectively" (H. W. Chesbrough, 2006). Gassman, Enkel, and Chesbrough identity nine research perspectives within the *open innovation* field and two of them directly specify *open source* and *proximity*. They discuss the *spatial perspective* as describing the absorptive capacity of the area, making the point that through *proximity* a firm is able to tap people with knowledge and world-class competencies without having to employ them (W. M. Cohen & Levinthal, 1990). The *cultural perspective* stresses that "the innovation process starts with a mindset" and that for *open innovation* to work, the culture needs to value outside competence (Gassmann et al., 2010). In the case of *open innovation*, valuing outside competence refers to using an *open source* approach. At the firm level, that outside competence can also apply to the organization, thus Kanter's description of the "open door policy" that reflects an open communication culture encouraging people at all levels inside an organization to contribute to innovation. This openness can translate as having a policy against closed meetings and even re-organizing the office space in order to eliminate private offices (Kanter, 2000).

By the end of the 1990s, the software development industry had been using *open source* as a business approach for at least two decades; during this time the approach spread to other industries involved with innovation. Chesborough and other scholars noticed this trend and began researching, writing case studies, and publishing about *open innovation* and *open source* business models as a valid business premise, which resulted in *open source* quickly becoming a popular topic in academic research (Baldwin & von Hippel,

2011; H. Chesbrough, 2006; Chesbrough, 2013; Chesbrough & Bogers, 2014; H. W. Chesbrough, 2006; Dedrick & West, 2003; DiBona & Ockman, 1999; Goldman & Gabriel, 2005; Hippel & Krogh, 2003; Lee, Olson, & Trimi, 2012; Lerner & Triole, 2000; Steele, 2012; Von Krogh & Von Hippel, 2006). In an analysis of the growth of *open innovation* publications, it was determined that by 2010, there were already 306 core publications with over 10,000 references (Raasch, Lee, Spaeth, & Herstatt, 2013).

Industrial districts.

Alfred Marshall (1842-1924) provided only a conceptualization of the industrial district through his books on economics; his study lacked a formal definition, but his narratives were descriptive (Marshall, 1920). A definition eventually emerged in the 1970s when Giacomo Becattini reinterpreted Marshall's work in Italian (Becattini, 1975). The Marshallian industrial district became a global academic topic when it was translated into English in 1989 (Goodman, Bamford, & Saynor, 1989). Marshall had a unique appreciation for the industrial district as a product of the people and the culture; the people and the firms were inseparable as a unit of analysis. The evidence of the concept of *open source* is in the original Marshall text from 1920 and can be gleaned from his description:

> When then an industry has once chosen a locality for itself, it is likely to stay there long: so great are the advantages which people following the same skilled trade get from near neighbourhood to one another. The mysteries of the trade become no mysteries; but are as it were in the air, and children learn many of them unconsciously. (Marshall, 1920)

This idea that the local culture embodied the industrial product to such a degree that they were inseparable concepts, reflects an organic and unconscious *open source* approach. An empirical study by Muscio revealed not only that firms located inside an industrial

176

district more innovative than those located outside, but he also explained why:

> Innovative firms outside districts follow a more 'linear' process of innovation, relying on internal R&D activities and on external research institutions. Firms inside districts, on the other hand, have proved to follow a different, more 'collective' pattern of innovation development, relying on the access to the local community of firms and people. (A. Muscio, 2006)

The *collective* pattern Muscio describes seems indicative of a communal type of *open innovation* or an *open source* approach to sharing tacit knowledge.

Innovation districts and community context.

An innovation district is referred to as 'hyper-local placemaking' (Rainwater, 2014). It reflected a discreet geographical area marked for a specific type of development focused on placemaking to foster innovation. Mixed-use and residential components may just seem like just attributes of a conventional development, but what is remarkably different about innovation districts, is that these components of life are embedded into the urban fabric, which contains a real community of people and a distinctive culture. The earliest industrial district literature reflects that socio-economic understanding, but clusters and research parks literature eliminates the concept of community from their context as a unit of analysis (Sforzi, 2015). This reuniting of an innovation industry with its community is why *open source* is ubiquitous in these early reports on innovation district development.

For example, recent sources demonstrate the theme of *open source* in innovation districts. "The organic results of profound economic and demographic forces are altering how we live and work. The growing application of 'open innovation'—where companies work with other firms, inventors, and researchers to generate new ideas and bring them to market—has revalued proximity, density, and other attributes of cities" (B. a. W. Katz, Julie, 2014). In an early publication, one of those authors referred to innovation districts as "crowd sourced rather than closed sourced" referring to the collaborative nature of crowd-sourced resources rather than closed-source code (B. Katz & Bradley, 2013). Crowdsourcing refers to the method of gathering input or resources from a group of people responding to an open invitation to collaborate or participate in a goal.

Innovation district stakeholders learn from each other. Sometimes they send delegations from one city to visit a more established innovation district in another city; other times academic researchers develop case studies to mine for lessons. In a case study developed by Canadian researchers, lessons were gleaned from the planning process of the @22 innovation district in Barcelona and applied them to the future innovation district in Montreal, now known as Quartier de l'Innovation (Battaglia & Tremblay, 2012). The authors suggest the Montreal Innovation district, the QI, "could be considered as an emerging innovation cluster that is developing through a "new model" of clustering and technopolitan approach" (Battaglia & Tremblay, 2012). They also cited the "denounced lack of public participation" from local community and social groups as an approach that limited the regeneration efforts of @22 (Battaglia & Tremblay, 2012). This evidence is an indication that by *not* applying the *open source* approach during the planning of an innovation district, results were less optimum. The case study quotes a stakeholder as specifying that "we must create an open innovation system in the territory rather than a closed innovation system in order to avoid the creation of a technological ghetto"

178

(Battaglia & Tremblay, 2012). Clearly, the lesson stands that open innovation should be an imperative in innovation district developments.

The ecosystem.

Internal to every place of innovation, especially the collaboratively formed innovation district, is a designed *innovation ecosystem*. One publication about *innovation ecosystems* attempted to wade through the literature "of loose and inconsistent use of the term" to build an understanding of how an *innovation ecosystem* differed from a science park, innovation cluster, and regional innovation systems (Oh et al., 2016). The authors listed as: "open innovation: the borrowing of licensing, open-sourcing, crowd sourcing, and alliances that allows ideas from diverse sources to be combined into new products and services" as a differentiating feature unique to the *innovation ecosystem* (Oh et al., 2016).

The university: experimental, spillovers, & corporate co-location.

Experimental.

Innovative thinking is not just in the domain of the private sector; it has long been part of the higher education system. In 1927, the University of Wisconsin allowed an audacious experiment, in what we would now recognize as *open source* information exchange, by allowing their Experimental College to design a "two-year undergraduate curriculum that integrated discrete school subjects to help students construct for themselves a 'scheme of reference' adequate to future study or life beyond college" (Meiklejohn, 1932). The curriculum eliminated required class attendance and one-way directional teaching in the form of lectures and examinations. It substituted reading lists, papers, and weekly individual tutorial conferences with professors. Students were treated as "adults and fellow inquirers with the faculty," which is a reflection of the *open source* approach manifesting on the fringe of this experimental effort in higher education.

The experimental college movement gradually spread to a handful of universities after Meiklejohn's program ceased, but enjoyed a robust growth after WWII from 1945-1965, during which time the interdisciplinary, self-created programs of study flourished (Stickler, 1964). In 1965, ten representatives from universities with experimental colleges held a conference in Florida to share experiences and plan growth for the movement; their proceedings were published as a book to inform the larger experimental college movement advocates (Stickler, 1964). Each program evolved from within its college, as the entire movement peaked in the late 1960s and early 1970s (Kliewer, 1999). In a review of a dozen case studies of existing experimental colleges, Kliewer identified that an egalitarian approach was consistently present among them (Kliewer, 1999). By definition, egalitarianism operates on the principle that all people are equal and deserve equal rights and equal voice. Where egalitarianism is present, the spirit of *open source* can flourish.

Spillovers.

Fast-forward to the twenty-first century and *spillovers* between the university and its host city have become recognized as a vital component of innovation in a region. Allison and Eversole cite two drivers of change have emerged that impact the way the university is traditionally structured. First, the shift toward networks of *open innovation* and, second, the rise of user-generated innovation and demand-driven solutions; the merging of these two drivers creates wholly new opportunities (Allison & Eversole, 2008). Though that paper is based on regional universities in the Australian higher education system, it offers applicable insights into university engagement with knowledge systems. As with the authors from the United Kingdom, South Africa, Italy, and the United States, this Australian paper also demonstrates that the *open source* approach is being acknowledged globally as a phenomenon in innovation.

Variation in a city's innovation capacities depends as much on collaboration between agents and their ability to mobilize assets, as on the ability to create and diffuse new knowledge (Simmie & Hart, 1999). For example, creativity capability occurs when entrepreneurs are able to get non-redundant information. A social structure can be enabling or constraining, so introducing multiple networks of diversity balances the tensions from conformity, allowing the inventor to glean new insights without compromising their innovative propensity (Ruef, 2002). These networks of diversity are how the *open source* information is gathered.

An example of spillover within the university itself are those universities that reorganize their departments around thematic issues, such as Arizona State University, or eliminate their department structure altogether and reorganize around a single imperative such as Unity College in Maine did when they organized around sustainability as a overarching theme. Building interdisciplinary capacity has been an ongoing effort for several decades in higher education, though the structure of the university budgeting and discipline-based reward systems frustrates the efforts (Klein, 1990a).

Corporate Co-location.

Innovative places reflect that diversity is valued and integral to the innovation process. Corporate co-location that mixes faculty and students was documented in the first experimental college effort in 1927 (Meiklejohn, 1932). The University of Florida in Gainsville claims to be the first in the nation to embed an entrepreneurial-based academic community, Infinity Hall, inside their 40-acre Innovation Square mixed-use development (Ed, 2016).

Corporate co-location also refers to corporations being physically occupying office or commercial lab space in academic buildings. Innovation campuses house industry partners in their academic buildings to facilitate relationships that result in *open source* interactions and knowledge spillover, both of which benefit the faculty and students as well as the industry partner. Centennial Campus at NC State in Raleigh, North Carolina, houses 60 industry partners ("Vision 2034," 2015). In New York City, the new building currently being constructed on Cornell Tech's new campus on Roosevelt Island called CoLo, a 189,000 square foot academic building designed with flex space to house industry partners. In 2017, plane manufacturer Airbus is moving their 400 employees in the U.S. engineering center from downtown Wichita to a new corporate co-location building under construction at the Innovation Campus at Wichita State University (Heck, 2015).

Iterative Process Found in Places of Innovation

Ironically, the *iterative process* is so closely associated with innovation that identifying evidence of it in these various places was the most challenging of all 21 findings. Longevity of an industry or community suggests adaptations have occurred. Adaptation can be thought of as an attempt to survive or improve, therefore adaptation is iteration. The *iterative process* was identified as an inherent characteristic in the original industrial district concept, in the transformation of the current research park model, in the implementation of the recent innovation district developments, and in universities of yesteryear and today.

Industrial districts.

During the Industrial Revolution, new processes were rapidly being introduced to the manufacturing process, and the public's hunger for new and suddenly affordable products seemed insatiable. This high demand created opportunities to redesign products and iterate new and better versions of products because the demand was stimulated by lower prices due to mass production techniques. Integral to this process was the organic formation of the early industrial districts in response to this new demand. Alfred Marshall observed the iterations from within the communities of industrial districts in 1920:

> Good work is rightly appreciated, inventions and improvements in machinery, in processes and the general organization of the business have their merits promptly discussed; if one man starts a new idea it is taken up by others and combined with suggestions of their own; and thus becomes the source of yet more new ideas. (Marshall, 1920)

Authors agree that modern day empirical research on the link between industrial districts and innovation is one of the least common themes in the literature on industrial districts (Boix & Trullén, 2010; A. Muscio, 2006). One of the first articles to attempt to measure and identify the determinants of the 'I-district effect' on patent generation was published in 2010 (Boix & Trullén, 2010). Industrial districts have the intrinsic feature of "continuously innovating" and that has been a basis for the iterations that led to the community's continuous adaptation and, ultimately, their sustained growth (Piore & Sabel, 1984). A long history of industrial districts in Italy reflects that adaptability and innovativeness are further hallmarks of a communal capacity in Italy's industrial districts (Rogerson, 1993).

Research parks.

In a 2009 report about the next twenty years of technology-led economic development, it was predicted that the role of the research park would change rapidly and would "shift from a single research park model to investing in entire innovation zones" (Townsend et al., 2009). This included the revitalization of mixed-use development to create an environment to attract talent, but also encourage innovation (Townsend et al., 2009). These off-campus investments have already begun with Syracuse University investing $13.8 million in support of their Connective Corridor in a blighted area known as the Near West Side; the University of Cincinnati spending $150 million in community development; and the University of Pennsylvania spending $100 million on redevelopment (Porter, 2010). Townsend predicts the research park model will undergo a major reinvention by 2030 and predicts it is likely that the terms *research park* and *science park* will fade from the vocabulary of economic development (Townsend et al.,

2009). Aging research parks are considering how to transform themselves into 'urbanized suburban' research parks (B. a. W. Katz, Julie, 2014). The Research Triangle Park in North Carolina issued a redevelopment plan in 2012 that begins with a $50 million development on 50 acres. The first phase of the plan includes public spaces, a dog park, a sculpture garden, and a 5,000-seat amphitheater. Future development will include two hotels, two corporate office towers, several hundred apartments, and 300,000 square feet of retail in the center of the development (Bracken, 2015). These new research parks represent a major change in how they position themselves in the competitive landscape. As Townsend notes, "In these new environments, managing the activity and flow of people and ideas so they make connections will require a very different set of skills from the typical research park manager" (Townsend et al., 2009).

Innovation districts.

Cities are faced with iterating new versions of themselves. Leon published a report looking at the impact of immigration of knowledge workers on Barcelona and @22; he stresses iteration by stating, "cities that do not have the diversity of skills to re-invent themselves when faced with the industry, technological or marketplace changes will atrophy, lacking the human capital to generate new industries and employment and failing to attract new firms and direct investment" (Leon, 2008). He also confirms the spillover benefits, "for the city to benefit, it is essential to capture the knowledge and economic spill-overs into local firms and institutions and do so by pro-actively engaging both local and new international communities" (Leon, 2008). Spillover knowledge is essentially a surplus of knowledge that gets 'recycled' by other innovators who have a use for it. Spillovers fit neatly into the *open source* approach of freely sharing information in a culture where it is appreciated and expected.

Innovation districts are just one strategy a city can use to realign local resources and energy on a long-term economic development engine. Innovation district developments have a long planning phase because of the collaborative process that typically involves a city, a college, and a corporation, but in reality can potentially involve the regional governance agencies, vocational education, foundations, community agencies, private developers, educational partners K-12, etc. (Glenn, 2016). The variety of stakeholders—each having different goals and varying timetables—brings complexity to the collaborative process. But through that quality of complexity, an opportunity for iteration emerges. Additionally, the expense of the development means these communities are not built overnight, but have an opportunity to grow in a structured way as finances allow. The final form of the innovation district may take a decade or two to fully develop, which means it will go through multiple iterations until it finds what works for a particular context. The mix of initial partnerships attracts new entrants into the innovation district making evolution in general inevitable, but still specifically unpredictable.

University.

Though higher education in the United States had been in constant evolution since its founding, higher education had also become entrenched in its methods that were increasingly seen at odds with the rapidly changing world at turn of the twentieth century (Meiklejohn, 1932). The University of Wisconsin president gave permission for the leading thinkers of the faculty to invent a new model of liberal education for the first two years at the university. They were also empowered to make adjustments—to iterate. The director and faculty advisers learned by trial and error. They were running a social experiment and making observations on what worked and what did not. They dutifully recorded their unguarded observations and humble commentary in a series of reports that John Meiklejohn included in his 1932 book *The Experimental College*, thus leaving a trail of evidence of their determination to continuously iterate throughout their experiment.

The case of restructuring at Arizona State University is an example of iterating the mission and organization to support the growth of the university (Crow, 2009). The New American University is President Michael Crow's reconceptualization of how higher education will cope with demographic changes among the student body and they will develop interdisciplinary problem-solving approaches to modern complex problems. Some applaud the effort, while others deride it. It will likely take decades before the ASU approach can be fully evaluated.

Proximity Found in Places of Innovation

Proximity is a pervasive theme in economic geography research covering *communities of innovators* found in industrial districts, clusters, research parks, innovation districts, and universities. Innovation emerges out of a context, and the effects of *proximity* help explain that context. The evidence presented on *proximity* regarding innovation falls into three categories: *physical proximity*, *functional proximity*, and *emotional proximity*. The *physical proximity* describes the benefits of density that allow for spillovers in a knowledge system and for serendipitous encounters to occur. *Functional proximity* captures the intangible tacit knowledge unique to each place and anchored in local communities. *Emotional proximity* provides the forum for frequency of personal interactions that build trust over time.

Physical proximity.

The literature on industrial districts postulates that shared competencies within a geographic area are easily shared and distributed, yet Camison used empirical findings to assert that "location within an industrial district is not sufficient to favour the generation of innovation and, in fact, is subject to many variables" (Camisón & Villar - López, 2012). A counter to that notion comes from Audretsch and Feldman, who published research that when firms were located in an industrial district and in close *proximity* to knowledge sources, they innovate faster (Audretsch & Feldman, 1996). Muscio found that the *proximity* of the university to an industrial district facilities knowledge transfers and leads to increased private funding through industry collaboration (A. Muscio, Quaglione, D., & Scarpinato, M. , 2012). In an analysis of clusters, a statistical association was found between innovation and cluster strength after analyzing 248 manufacturing firms (Baptista & Swann, 1998). They found the strength of the cluster as measured by own-sector employment is correlated to innovation activity, but they had no explanations for the underlying mechanisms.

Innovation district development emphasizes physical *proximity* by using mass transportation to delineate boundaries and focus density within a specific area (Glenn, 2016). In a 2013 thesis about the role of the *third places* in the development of Boston's Kendall Square, it was postulated that the absence of attention to innovation in dense urban areas was likely due to the focus on the larger regional scale, such as the scale established by Silicon Valley, which is the dominant model of *innovation ecosystems* (Kim, 2013). The other thesis on innovation districts analyzed transit travel times between vital components of the *innovation ecosystem* in Boston to establish the role of *proximity* to the Kendall Square area and how those components could impact the spread of the innovation district concept to adjacent neighborhoods. These components consisted of colleges, universities, incubators, accelerators, and training programs in a radius around Kendall Square.

An innovation district operates on the premise that an *innovation ecosystem* needs to be physically identified on a map and in close *proximity* to the end users, the entrepreneurs. The authors of the @22 Barcelona-Montreal case study (discussed in the previous *open source* section) refer several times to the importance of spatial integration that concentrates firms and institutions for closer geographical *proximity*, citing the concept that "relational proximity fosters knowledge exchanges" (Battaglia & Tremblay, 2012).

In the wake of books such as *Distance is Dead* and *The World is Flat* that decried the value of physical presence, others now suggest *proximity* has to be reconsidered as important and revalued in the context of what is referred to as the 'innovation generation'. For example, "The growing application of open innovation – where companies work with other firms' inventors and researchers to generate new ideas and bring them to market - has revalued proximity density and other attributes of cities" (B. Katz, 2015).

The ease of contact offered by *physical proximity* is obvious: "Knowledge transfer between highly skilled people happens more easily in cities because the sheer density and concentration of economic players in large cities offer multiple opportunities for contact interactions and knowledge circulation" (Orlando & Verba, 2005). The ease of access to information is also recognized: "Proximity to the source of the research is important in influencing the success with which knowledge generated in the research laboratory is transferred to firms for commercial exploitation" (Bramwell et al., 2012). In looking for firm-university innovation linkages in Great Britain, researchers found the value of locating close to a university depended on industry *proximity*. Chemical and pharmaceutical firms benefited the most from knowledge spillovers when located near universities (Abramovsky & Simpson, 2011).

Functional proximity.

Alfred Marshall alludes to tacit knowledge in the early industrial district, an observation that he describes as the expertise being 'in the air': "The mysteries of the trade become no mysteries; but are as it were in the air, and children learn many of them unconsciously" (Marshall, 1920). This early recognition of the value of an intangible

culture to the local industry became a concept validated by subsequent researchers and

studied intensely:

> Tacit knowledge, on the other hand, cannot cope with distance. It can only be
> copied by means of observation, practice, or learning and since it is encoded in
> human beings and their daily behavior; it can only be transmitted via face-to-face
> contacts. The role of tacit knowledge in the co-localization of research activities
> and innovative enterprises is clearly evident. It is only within specially anchored
> communities that tacit knowledge can really be transmitted and transferred. (van
> Oort, Burger, & Raspe, 2008)

Tacit knowledge requires local contact, whereas codified knowledge can spread globally.

Tacit requires transfer from person to person and seems to defy being written in an

instructional form. Whereas codified knowledge is more straightforward and prescriptive.

As Boix & Trullén have seen in industrial districts, "Codified knowledge mostly refers to

scientific and technical knowledge compiled in codes that can be transmitted and learned

by means of the usual mechanisms of communications and formal education, and does

not need the experience of other people or a precise context" (Boix & Trullén, 2010). In a

competitive world, tacit knowledge is what gives *place* an advantage that is not easily

replicated. It is hard—if not impossible—to export an entire culture to a different country

that is the 'low-bidder'.

Knowledge in innovation communities tends to be tacit until the parts that can be

captured by printed word become codified. How much knowledge can be captured varies

based on what type it is. This process occurs as the need for information dissemination

increases: "The more frequently cited explanations for this proximity effect is the need to

gain access to tacit knowledge or at least knowledge that is not yet published in scientific papers" (Bramwell et al., 2012). This is why co-location of corporations within academic buildings on campus is key to participating in the knowledge spillover. There is more research generated than will ever be published, especially considering experiments that fail are traditionally not submitted and not published but those failed experiments provide invaluable information for a corporation. Innovation scholars recognize, "The more codified the knowledge involved, the less space-sensitive these processes tend to be. When the knowledge involved is diffuse and tacit, the argument is that such interaction and exchange is dependent on spatial proximity between the actors involved" (Bathelt, Malmberg, & Maskell, 2004). *Proximity* facilitates tacit knowledge flowing from a culture or institution to others through repetitive exposures. This spillover is a function of *proximity*. There is a personal interaction required for *proximity* to be of benefit, "firms located nearby universities are more likely to benefit from knowledge externalities from academia, as spatial proximity facilitates the interactions and face-to-face contacts necessary for the transmission of the tacit component of knowledge" (D'Este, Guy, & Iammarino, 2012). Because so much more information is conveyed unconsciously when delivered in person, "traditional face-to-face contacts remain an important condition for the generation and exchange of non-standardized and complex knowledge" (van Oort et al., 2008). The type of tacit knowledge is unique to each place based on a host of factors: traditions, culture, rituals, local knowledge, social networks, and intellectual resources; these factors differ from city to city and university to university. Innovation often has a geographical or social 'stickiness' to it because it can draw on "combinations of scientific knowledge, technical skill, and tacit knowledge that is place-specific" (Townsend et al., 2009).

Another measure of innovation activity is patent intensity. Because the knowledge flow is so hard to measure, patent citations are used as a proxy of innovation activity (Jaffe, Trajtenberg, & Henderson, 1993). Research showed that an employment density of 2,200 jobs per square mile generated the highest patent intensity (Carlino, Chatterjee, & Hunt, 2007). Beyond that density, there was a point of diminishing returns. Density, which is a physical measure of *proximity*, plays a role in creating a flow of ideas that results in patents, which reflect innovation (Carlino et al., 2007). Carlino both acknowledges and wrestles with the theory of *proximity*, "geographic concentration of people and jobs in cities facilitates the spread of tacit knowledge. While that mechanism is not well identified in theory, the underlying idea articulated in the market is that the geographic proximity created by density facilities the exchange of information among workers and firms" (Carlino et al., 2007).

Emotional proximity.

Proximity alone does not lead to information exchange; that requires trust be earned and built through frequent interactions. The relationship can be maintained through email correspondence and phone communication once the initial relationship is established by interactions in person. Researchers found that "exploration—establishing new connections among people—is an excellent predictor of innovation and creative output" (Pentland, 2013). Such 'scouts' who explore new relationships have an opportunity to vet those new contacts based on trustworthiness. Echoing the intrinsic value of personal relationships, "rich channels of communication, particularly face-to-face interaction, matter much more than electronic communication channels. In other words, e-mail can never fully replace meetings and conversations" (Pentland, 2013).

Corporate co-location provides this face-to face interaction. An innovation campus that incorporates the co-location of firms and academic departments in the same building radically advances the notion of integrating industry and academe. "Building on interaction is the foremost principle" because the campus can be designed to facilitate formal and informal contacts (Greiger, 2004). These face-to-face interactions are colorfully described by innovation district stakeholders as 'happy collisions' or 'creative collisions' and raised to an art form as a 'choreography of collisions.'

Long before co-location was an innovation strategy between firms and universities, faculty and students were collocated to foster stronger relationships. At The Experimental College within the University of Wisconsin (1927-1935), a single dormitory was designated to provide student living quarters, professors' offices, and classroom space. The goal was to "dissolve the distinction between academic study and college life" (Meiklejohn, 1932). This provided a great deal of "camaraderie of spirit" and *proximity* to the faculty was considered a "central formative factor" in a student's development (Meiklejohn, 1932).

Interestingly, another by-product of this *proximity* created a deeper appreciation and acceptance of the differences between students at the University of Wisconsin in the 1920s and 1930s. Students of different ethnic groups learned to work together. Students from different religions enjoyed an unusual unguarded exchange of perspectives. In *The*

Experimental College, author John Meiklejohn quotes Cardinal Newman in his book, *The Idea of a University,* as observing that a tight-knit community "gives birth to a living teaching, which in the course of time will take the shape of a self-perpetuating tradition" (Meiklejohn, 1932). From an *emotional proximity* perspective, this quality of a *living teaching* that creates a culture and reputation would seem to be a tremendously worthy aspiration for any university or corporation.

Not quite a hundred years later, leading thinkers of the faculty are again considering how to catalyze 'innovation in place' within higher education, this time for Australia's regional universities. Allison and Eversole go beyond listing the obvious physical assets such as distinctive landscapes and livable environments to stress how the role of knowledge comes into play for innovation (Allison & Eversole, 2008). Such locally-specific knowledge creates a distinct set of attributes, which is important because "these attributes, rooted in *relationships of proximity,* cannot be easily or quickly replicated—hence generating a competitive advantage" (Amin, 2004). The key word in this quote is not just a reference to *proximity*, but the 'relationships of *proximity*' because a trusted relationship is the end game of *proximity* when trying to foster innovation.

Diffusion of Sustainability-oriented Innovations

A historical review of the greening in the built environment, the greening of business, and the greening of higher education indicates that there is already an ongoing response to innovations around sustainability. The growth of such innovations supporting sustainability in these industries provides a foundation of real world evidence, which justifies the need for recommendations for the implications of *open source*, *iterative process*, and *proximity*.

The Greening of the Built Environment

In the United States, the built environment uses about 30% of the energy and transportation uses about 40% of the energy ("Green Building," 2016). Given these significant uses, it is understandable that the environmental movement of the 1970s would quickly focus on the built environment. The seeds of the green building movement took root in the 1970s in the *back-to-the-land* movement and in the architecture community. Ecological construction and ecological living books proliferated. *The Mother Earth News* magazine was launched in 1970 and remains in print today. *The Good Life* was a how-to account of Scott and Helen Nearings' experiences as modern homesteaders in rural Vermont during the Great Depression era. *The Good Life* subsequently captured the imagination of the young adults of the 1960s who were disillusioned and seeking a more Earth-friendly path in life. It is considered the premier manual by the back-to-the-land movement advocates then and now. *The Toilet Papers* by Sim Van der Ryn inspired his architecture students at Berkeley and beyond. Van der Ryn's 1997 book written with

Stuart Cohen, *Ecological Design,* informed generations of architecture students and practitioners. *Regenerative Design for Sustainable Development* by John Tillman Lyle in 1996 is considered another must-read for ecological designers. Lyle initiated an ecovillage research station at Cal Poly-Pomona (K. Brown, 2015). In the past decade biomimicry books—the application of efficient designs developed in nature—have found their way into the design of sustainable building products. Tools and resources now proliferate mainstream in support of green design and green building.

One of the earliest green building programs emerged in Austin, Texas in 1990. High-performance green buildings are defined as: "facilities designed, built, operated, renovated, and disposed of using ecological principles for purpose of promoting occupant health and resource efficiency, plus minimizing the impacts of the built environment on the natural environment" (Kibert, 2004). The United States Green Building Council (USGBC) was formed in 1993 for the purpose of promoting a green building rating system for commercial construction known as LEED (Leadership in Energy and Environmental Design). In 1999, there were just a few buildings that were LEED certified, but by 2003 over 400 buildings were certified. As of 2013, there were 186,000 professionals who held LEED credentials (USGBC, 2013). As of 2015, more than 26,600 projects had been LEED-certified while another 42,000 projects have pending applications.

The Declaration of Interdependence for a Sustainable Future report was produced in 1993 as a joint effort of the International Union of Architects (IUA) and the American

Institute of Architects (AIA). Ed Mazria, of Santa Fe—a green architect known for a seminal book on solar energy—developed the 2030 Challenge in 2002, and launched it in 2006 under the banner of the American Institute of Architects as an industry challenge. The AIA 2030 Challenge called for all new buildings to be carbon neutral by 2030. This means that the energy used to operate the building cannot not be provided by a fossil fuel that produces CO_2 emissions. Along a similar premise, the zero-energy building concept (ZEB)—buildings that generate all their energy use from on-site sources—has been in development for decades in all parts of the world. In the United States, the Department of Energy (DOE) has researched and promoted ZEB for commercial construction and ZEH (zero energy homes) for residential construction. To bring clarity to the industry, the DOE recently issued a definition of ZEB as "an energy-efficient building where, on a source energy basis, the actual annual delivered energy is less than or equal to the on-site renewable exported energy" (DOE, 2015).

The next evolution in green building and zero-energy construction came in the form of a holistic rating system called the Living Building Challenge (LBC) invented in 2006 by architects Jason McLennan and Bob Berkebile. It comprised 20 imperative benchmarks in seven areas: site, water, energy, health, materials, equity, and beauty. A new and broader scale, The Living Community Challenge rating system, is scheduled to launch in 2016. These efforts are resulting in high profile 'signature projects' at this stage of growth. Regardless if the LBC develops into a widely accepted industry standard or not, it is generating an evolution in ecological design. LBC serves as a harbinger of thought leaders intentions and signals the overall direction of the built environment industry.

In addition to the construction industry and architecture industry response to global environment challenges, the planning industry component of the built environment has contributed solutions to address environmental resource scarcity over the past few decades. From the practitioner realm, sustainable planning emerged under the various names such as New Urbanism, Tradition Neighborhood Development (TND), Transit-Oriented Development (TOD), Conservation Design for subdivisions, and Agricultural Urbanism. Transitions Towns is a grassroots community effort to move a community to becoming fossil fuel free. The movement emerged in 2006 in just one city in the United Kingdom; but by 2014, over 1,100 initiatives were registered worldwide (Aiken, 2012). As for the landscape architecture industry, they adopted the Sustainable Sites Initiative in 2007. Sustainable Sites is a nationally acclaimed, evaluative tool to guide site design in consideration of twelve ecosystem services provided by a given landscape.

The industries within the built environment have long used the terms efficiency, sustainability, and placemaking, but in the past decades—in the wake of super-storm hurricanes and economic shockwaves—the term 'resiliency' has become the new buzzword (Mazur, 2015). Though the professionals in the built environment are keenly aware of the environmental challenges, the education and the ethic has been slower to migrate to policymakers and investors; innovative environmental approaches have been introduced, but are not yet institutionalized broadly in our society (Sharifi, 2016). Nevertheless, some diffusion has occurred. For example, the Presidential Executive order #13423 signed in January 2007 addressed federal goals for sustainable design and high performance buildings and The Energy Independence and Security Act of 2007 included requirements for high-performance green federal buildings ("Green Building," 2016).

The fact that solutions have been generated by the professional societies from the built environment over the past five decades—architects, planners, construction firms, and governance agencies—signals a shift in awareness that validates how an industry responsible for massive energy use has responded with better design, policy changes, professional development, and philosophical changes. Yet, what is equally amazing is how slow progress truly has been over the past 50 years, even with humanity's existence in peril. As early as the 1960s, academics were calling for a policy response to degradation and societal decline; they publicly questioned if man's evolution of responsibility had keep pace with the evolution of technology (Winthrop, 1963). Innovations around sustainability abound in the built environment, but is the diffusion fast enough to regenerate the environmental damage and move closer to a sustainable equilibrium?

The Greening of Business

Where was the ethos of sustainable business birthed? Likely, it was in the hearts and minds of entrepreneurs operating in the new culture after the environmental movement in the 1970s. Outdoor clothing company Patagonia positions itself as being a champion of the environment since its inception in 1973. Patagonia continues to lead by example; their mission statements is "build the best product, cause no unnecessary harm, use business to inspire and implement solutions to the environmental crisis" (Patagonia, 2015). As the 'conscious capitalism' trend grew, it became recognized and legitimatized by Paul Hawken in his 1993 book *Ecology of Commerce: Declaration of Sustainability*. This potent book serendipitously fell into the hands of Ray Anderson, founder and CEO of a billion dollar carpet manufacturing company, who was reluctantly pondering how his firm's environmental policy could go 'beyond compliance'. Later, as he recounted reading the chapter titled "The Death of Birth," he described that phrase as a "spear in his chest." Ray Anderson became known as the Greenest CEO in America. In the year 2000, he put his company and its products—derived from fossil fuels—on a twenty-year trek to climb what he coined, Mount Sustainability; it was an apt metaphor that acknowledged the difficulties ahead and implied what kind of determined leadership would be required to succeed. Ray Anderson was an early and perpetual force within the green building industry and became a key consultant to Walmart prior to their public entry into corporate social responsibility.

The consulting relationship between Ray Anderson and Walmart led Lee Scott, the CEO of Walmart, to take a courageous and audacious position around sustainability. An unforeseen event led to Walmart enjoying the experience of being viewed, for the first time, as a community hero. After Hurricane Katrina devastated the Gulf Coast in August 2005, Walmart become a vital partner to the region by using its logistical expertise to mobilize food and supplies to those in dire need in critically damaged areas. It was this experience that emboldened Lee Scott to reimagine Walmart being a leverage point for positive contributions in sustainability. He delivered a radical and powerful speech at the Walmart annual meeting in October 2005. Years later, Paul Hawken disclosed during his own speech at GreenBuild that it was he who had penned Walmart's *21st Century Leadership* speech that had sent shockwaves through the corporate world. Walmart began their sustainability operations by applying their world-class data management systems to develop sustainability metrics. In 2009, Walmart founded the Sustainability Consortium, co-administered by Arizona State University and the University of Arkansas; their purpose was to link scientific data to sustainable business operating decisions, thus driving innovation through a 'sustainability index' (Walmart, 2015).

Though sustainability-savvy entrepreneurs were practicing green business ethos for decades, it was John Elkington who captured this emerging trend by introducing the triple bottom line (TBL) terminology in his 1997 book *Cannibals with Forks* (Elkington, 1997). Recognizing that businesses needed a formal structure for explaining the triple bottom line metric, that same year a framework for creating corporate sustainability reports was created in Boston by Robert Massie and Allen White through their affiliations with the CERES (originally Coalition for Environmentally Responsible Economies) and Tellus organizations (Moore, 2012). The first Global Reporting Initiative (GRI) template was released in 2000. As of 2014, over 7,500 corporations worldwide voluntarily applied the framework to describe corporate social responsibility efforts (CSR). The success and adoption of the GRI framework led to a parallel development called 'integrated reporting' that links the legally mandated financial reporting with the voluntary CSR reporting. The Integrated reporting approach was bolstered by two Harvard professors who published a book on the topic and coined the term 'one report' (Eccles & Krzus, 2010). Currently, there is an effort by the International Integrated Reporting Council to create a global standard and expectation ("International Integrated Reporting Council ", 2016).

The greening of corporations created a demand for new levels of efficiency, product sourcing, and environmental responsibility. This demand created opportunities for 'lean manufacturing' processes designed to reduce waste. The demand also created a venue for the 'green chemistry' movement to evolve through an affiliation with higher education; the green chemistry approach looks for ways to reduce waste and reduce toxicity. Alliances evolved between the green building industry and the green chemistry community because both groups saw the potential for collaboration to increase indoor environmental air quality through better green materials. The sustainability ethic moved from the fringe of green entrepreneurs in the 1970s to the mainstream of both the corporate world and higher education.

The Greening of Higher Education Institutions

Teaching sustainability requires system thinking and system thinking requires interdisciplinary mindsets. Innovation in interdisciplinary efforts can be traced back through the development of higher education. One example is from 1927 when the University of Wisconsin hired John Meikeljohn to spearhead a novel two-year curriculum designed to create critical thinkers who would be prepared for advanced study (Meiklejohn, 1932). Meikeljohn's Experimental College sparked a wave of prototype efforts that culminated with hundreds of experimental college interpretations by the 1960s (Stickler, 1964). As of 1999, there were still over 300 in existence (Kliewer, 1999).

A trend related to the experimental college movement is the growth of interdisciplinary initiatives, which has experienced a prolonged renaissance beyond the scope of this discussion, but is well documented by scholars (Klein, 1990a, 1990b; Newell, 2001; Vincent, Danielson, & Santos, 2015).

Within this growth of interdisciplinarity is found a slew of new green degrees and sustainability science programs: ecological economics, green MBAs, green marketing, sustainable agriculture, sustainable architecture, among others (Delong & McDermott, 2013; Hart et al., 2015; Mayumi, 2002; Newell, 2001). As recently as 1987, not a single business school in the United States offered an environmental course (Galea, 2007). As of 2015, 46% of the top-100 MBA programs offered corporate social responsibility and sustainability programs (Hart et al., 2015). There are now a plethora of 'green' or 'sustainability' degrees including new doctoral degrees such as Sustainability, Sustainable Development, and Sustainability Science. Sustainability degree programs have grown from just 1 in 2006 to 141 in 2012 (Vincent, Bunn, & Stevens, 2012).Unity College in Maine recognizes the twenty-first century as the 'environmental century' and they developed the Environmental Citizen Curriculum as a common core for all their degrees ("Sustainability science a framework for the future," 2016).

In the past decade or so, new national organizations have arisen to reflect and compliment the growth of sustainability interests on campus. The Association for the Advancement of Sustainability in Higher Education (AASHE), Eco-reps, and Net Impact are all new national organizations on campus in support of the sustainability directive.

For a broader view that considers the growth of sustainable innovations on the facilities side as well as the curricular developments, Mary Whitney's dissertation looked at the interactions between the six most popular sustainability agreements at 1,400 universities in the United States (Whitney, 2014). The agreements were the Talloires Declaration, the Association for the Advancement of Sustainability in Higher Education (AASHE), the Sustainability Tracking, Rating & Assessment Systems (STARS), the American College and University Presidents Climate Commitment (ACUPCC), the International Sustainable Campuses Network (ISCN), and the Princeton Review Green Schools. Table 6 lists how many universities have signed these agreements and the year those agreements were launched. The Talloires agreement states their signatories on a worldwide basis, but the numbers reported by all other agencies apply to signatories in the United States.

Table 6. Sustainability Agreements in Higher Education in the United States

Agreement	Signatories	Year Established
Talloires	430 worldwide	1990
AASHE	783	2004
STARS (with AASHE)	247	2008
ACUPCC	400	2007
Princeton Green Schools	322	2007
ISCN	47	2009

With 1,400 out of roughly 1,800 universities committing to at least one kind of green initiative, Whitney considers this majority involvement as evidence of a social movement as well as an institutional movement. Citing a lack of leadership at the macro-level of the federal and state governments, Whitney explains that higher education institutions have pursued voluntary agreements that span campus operations, building construction, research, curricula, and outreach (Whitney, 2014). It has taken about a quarter of a century for 77% of the universities in the United States to enter into a public written agreement to address sustainability. Operationalizing these goals and promises is the next phase and will likely require additional innovations in practice and policy as well as public demand.

To summarize, evidence for sustainability in higher education has grown tremendously since the first Earth Day in 1970 and is rapidly becoming a ubiquitous topic on campus, even if its layered complexities are only loosely understood. Even with this penetration, society has barely begun to address all the *sustainability-oriented innovations* that will be required to live in a state of resource equilibrium with the natural resource base of the Earth. In chapter 5, the discussion turns to how innovation can be fostered using the findings of *open source*, *iterative process*, and *proximity* identified in these investigations of *communities of innovators*.

CHAPTER V

CONCLUSIONS

Underlying Problem

Virtually every product and every process in the world has to be reinvented in accordance with sustainability principles if the world is to reach a sustainable equilibrium (Z. Goldsmith, 2007). As Paul Hawken succinctly sums up in the classic business text *Ecology of Commerce,* we have a 'design flaw' (Hawken, 1993a). To the matter of *why* human beings, with all their intelligence, would even produce such a self-destructive design flaw in their approach to economic civilization is a topic of endless philosophical exploration. This realignment of the economy will require an unprecedented number of innovations to be designed and developed under a paradigm that is significantly different than the one that created the world in which we presently live. A call for an entirely new paradigm may seem radical, but actually it is not; this call is for a paradigm based on wisdom and compassion. It is a paradigm committed to environmental management, restorative mindsets, and regenerative principles. A major issue facing humanity is sustainability, the ability of the human species to prosperously endure for infinite generations given finite natural resources. In a sustainable world we, as stewards of the limited resources on Earth, will need to understand and respond to the interplay between the dynamic complex systems in which we live. This shift will require a significant change in the way we learn and collaborate as well as how we measure our holistic economic success—all which can result in a more balanced prosperity that is sustainable. As a knowledge society, it is logical to look to those who generate knowledge—the higher education institutions—to train future citizens who have the ability, propensity, and deep desire to create a culture of innovations for sustainability. To solve the conundrum of unsustainability that higher education institutions have inadvertently helped create, its leaders, in the immediate future, will be behooved to generate meaningful applied research and curriculum that contributes to the reversal of unsustainability with the goal toward sustainability. To establish a foundation for sustainability, we must shift away from the rote and algorithmic approach to learning that has been prevalent in twentieth century and move toward a more relevant heuristic model of learning that fosters innovative thinking (Pink, 2011). This begins with understanding the mechanisms that higher education leaders can use to systemically address curricular alignment with sustainability principles. Foundational mechanisms can scale up to apply to strategic long-term planning and scale down to impact classroom pedagogy.

How a problem is approached reflects the perspective or, quite literally, the thinking anchored in the values. Different thinking is needed to produce different results. To quote Albert Einstein—in the original context of failing to recognize the destructive capabilities of the atom bomb and 'drifting toward unparalleled catastrophe'—he simply points out, "a new type of thinking is essential if mankind is to survive and move toward higher levels" (Einstein, 1946). To realize the countless *sustainability-oriented innovations* necessary to operate our civilization—that is likely again 'drifting toward unparalleled catastrophe'—we need realistic, idealist, utopian thinking and realistic, idealist, innovative thinking that, when fused into one recognizable concept, is simply *sustainability thinking*.

Purpose of the Study

The study evolved from an effort to triangulate community, innovation, and sustainability to articulate the role of innovation in creating the desired state of civilization and the means to create it. The goal of this dissertation about *sustainability-oriented innovations* is to identify the mechanisms and environments common to various communities of innovators who lived and worked in places supportive of continuous improvement. The premise of this study is that by having better insights around the conditions that support innovation, higher education institutions can better leverage their communities and the uniqueness of their place so that they can becomes a specific kind of venue capable of creating new generations who will advance society—not just toward generic growth, but in pursuit of genuine development around sustainability intentions.

The focus is on the *approaches* used to generate innovation, not the degree to which these *intentional communities* or *places of innovation* were successes or failures. The reason for this focus on approach rather than outcome is that the metric of success regarding an *intentional community* is a subject of considerable debate among scholars who have concluded conventionally-defined success—such as longevity, societal contribution, or financial return—is not only undeterminable, but an undesirable goal (Aguilar, 2012; Fogarty, 1980; Kanter, 1972; J. Lockyer, 2009; Sargent, 2012; J. Wagner, 1985). As for the metric of success for *places of innovation*, that too was a difficult determination. Often the patent rate is used as a proxy for innovation activity, but a place such as an industrial district offers much more to the broader region besides economic returns. Sforzi explains that Becattini, the original Italian industrial district scholar, critically examined the positions of various economists by identifying the different ways industrial

districts could be analyzed; within this analysis was the consideration of satisfaction of

needs and the sense of belonging provided by an industrial district (Becattini, 1962;

Sforzi, 2015).

Research Question

The primary research question investigated was "What are the mechanisms historically used by *communities of innovators* as identified in *intentional communities* and in *places of innovation* that were used to approach their goals?" For implications, the research question was then tailored for the university setting "How can these mechanisms be applied to the environments created by higher education institutions so they can successfully fuel innovations that advance sustainability?"

Methodology

The research began as an inductive exploration and culminated with themes that offer an emerging theory that recognizes the underlying mechanisms used by innovators living and working within a community. Historical methodology is commonly used to reconstruct the past and—occasionally, used judiciously—to imagine potential futures. Process tracing was the systematic procedure employed, within the historical methodology, to analyze *intentional communities* and *places of innovation* for themes that were common to both. By using process tracing to recognize patterns occurring in the past, the research revealed a current convergence of each of the research fields. The result was the recognition that the innovation district is a rapidly emerging form of development. It presents itself as a hybrid of the past forms of development or settlement; the innovation district is an 'intentional place for a community of innovators'. The historical methodology can be credited with organizing the data in context of the past and in context of the future. The process tracing method, borrowed from political comparative politics, can be credited with evaluating each piece of evidence that provided a foundation for the generalization of the emerging theory generated from the research. The narratives interpreted and hypothesized provide a platform to discuss potential responses from higher education leaders regarding the socictal need for sustainability and the role of the university regarding *sustainability-oriented innovations*.

Summary of the Findings

This research illustrates the fusion of innovation and sustainability, and the potential evolution with the collaborative efforts originating in a community or a collective. When the two realms of innovation and sustainability overlap, it creates *sustainability-oriented innovation* as illustrated in Figure 3 (Jay & Gerard, 2015).

Figure 3. Overlap Creates Sustainability-Oriented Innovation

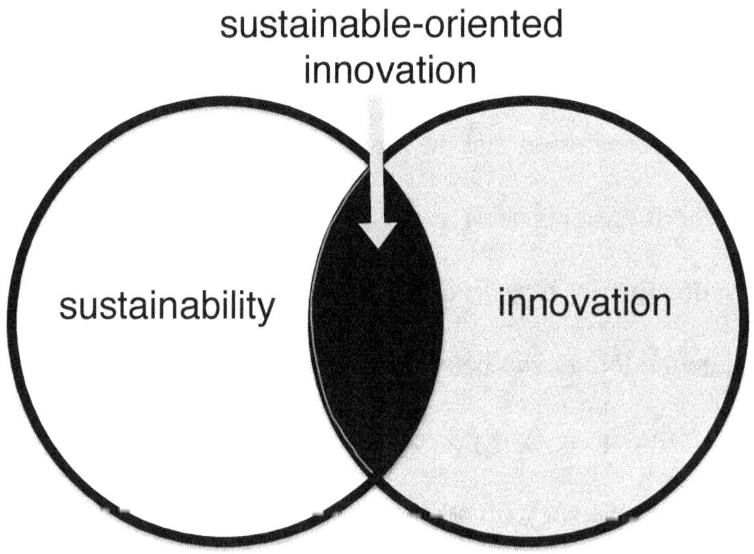

The overlap sliver of innovation and community creates innovation districts. The overlap sliver of sustainability and community creates ecovillages. Notice the community realm does not intersect with *sustainability-oriented innovation*.

The goal of a sustainable society would be to shift these three circles for maximum

overlap so that the global society is comprised of local communities who are participating

in the creation and implementation of *sustainability-oriented innovations* as illustrated in

Figure 4.

**Figure 4. Overlap of Sustainability-Oriented Innovation
from Community Collaboration**

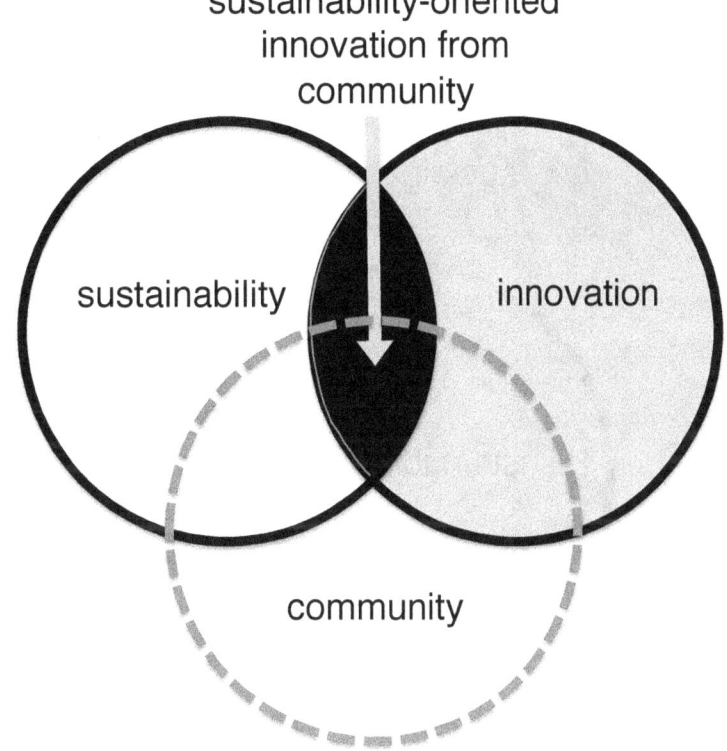

On occasion, ecovillage residents can shift their intention into the innovation realm to

create solutions that could be considered *sustainability-oriented innovations* resulting

from a collaborative effort from within a community. Innovation districts are just now

experimenting with sustainability overlays in the form of blending Eco-District protocols with their innovation district missions, but it is quite likely that this will also result in *sustainability-oriented innovations.* In order to identify emerging trends, it is important to notice minor movements such as these examples operating at the cusp of progress. Equally important is asking how to facilitate this desirable shift, thus the theory on mechanisms that foster sustainability-oriented innovation. Figure 5 recognizes these relationships that emerge from the overlap.

Figure 5. Overlap that Creates Communities of Innovators

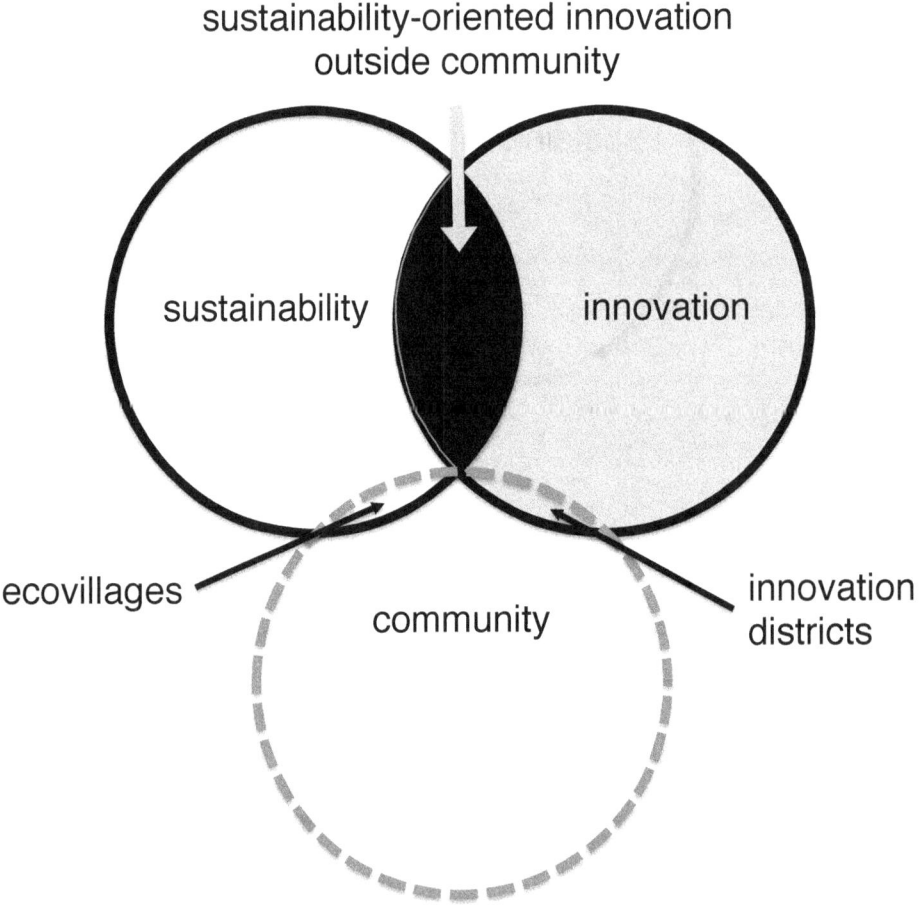

Throughout my decade-long quest to develop a deeper understanding of sustainability in practice, I encountered individuals and industries imbued with a curious underpinning of similar values that were reflected in the approaches they used to move toward sustainability goals. I engaged, participated, and tracked sustainability diffusion in three industries: the built environment, business, and higher education. As my research attention turned to how higher education could foster the innovations necessary to advance sustainability in practice, I found the same value-laden approaches, used by entities renowned for advanced sustainability, reflected in the two areas investigated in this dissertation: *intentional communities* and *places of innovation*. After extensive pattern searching and iteration, three themes became prevalent in these *communities of innovators*: *open source*, *iterative process*, and *proximity*. These themes comprised an emerging theory about fostering innovation; the acronym creating the term *OSipp*. The working hypothesis turned into an emerging theory, the *OSipp Theory*, as illustrated in Figure 6.

Figure 6. OSipp Theory: Three Mechanisms that Foster Innovation

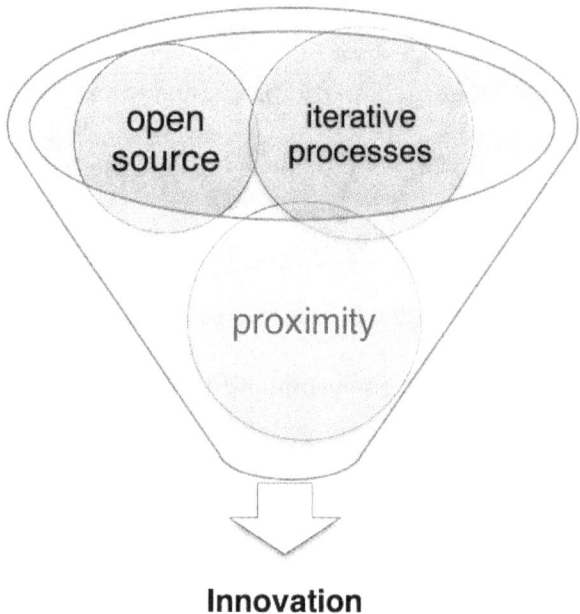

Innovation

To borrow the computer metaphor, the operating system (OS) is what executes the software programs just as the *OSipp Theory* comprises the approach by which *communities of innovators* pursue their goals. The *open source* concept, historically on the fringe of consciousness, had increasingly become a pervasive approach to organizing social innovations due to internet connectivity (Leadbetter, 2008). The business model of *open innovation* is a popular business philosophy that has gained significant momentum over the past decade (Chesbrough & Bogers, 2014). It has its roots in *open source* information exchange (Leadbetter, 2008). The *iterative process* approach originated in engineering and was popularized through private industry (such as IDEO the product design company) that used it as the basis of product design. Soon thereafter, the *iterative process* approach was institutionalized by the Stanford d.school where it was taught as a skill known as *design thinking* (Tischler, 2009). *Proximity*, as an academic observation, originated in the fields of economic geography, architecture, urban planning, landscape architecture, and innovation studies. The working definition of the *OSipp Theory* is:

> an approach intended to foster innovation by combining mutually reinforcing mechanisms of *open source* for information exchange used with an iterative approach to finding solutions which are utilized by a community of people who interact through a physical, functional, or emotional proximity.

A worthwhile theory goes beyond describing the mechanisms; it goes deeper to explore *why* these mechanisms function. The *OSipp Theory* is a problem-solving tool observable only through actions. It is driven by tacit values that are unknowable to the historian, yet indicative of the community values. I posit that a global sustainable equilibrium is achievable if—and only if—decisions are transparent and anchored in a sustainability ethic framework such as the Earth Charter.

Innovation, in the generic form, will lead to a civilization that is novel, flashy, entertaining, amazing, life-enhancing, life-saving even, but it will not lead to a just, equitable, or environmentally sustainable society. Generic innovation is positioned as the elixir for prosperity; "innovation has become a cure-all for economic woes" (Shearmur, 2012). This is why, throughout this dissertation, innovation has been steered toward *sustainability-oriented* innovation. The responsibility has been cast upon higher education institutions because they are the *one* set of institutions with the capacity to understand the intellectual and scientific basis of sustainability and they—not industry and not municipalities—are the major producers contributing to the *knowledge economy*.

Open Source

Open source is an approach of sharing information that is sometimes challenging to determine in these various *places of innovation,* because the previous research has focused heavily on the inputs and outputs for policy decisions rather than focusing on the individual places themselves or the holistic benefits to the greater community. In the nineteenth century when industrial districts were recognized as an academic research topic, Alfred Marshall described the local expertise found in industrial districts as 'in the air'—literally, saturating of the culture (Marshall, 1920). Marshall's early contribution to economic policy was to shift the traditional units of analysis of the individual firm to an intermediate unit of analysis based on local systems of production such as the industrial district, thus broader than the individual firm level (Tappi, 2001). This shift in research focus complicates the investigation for *open source* and requires a specialized discernment to tease it out.

The early recollections of the first science parks reflect the 'spirit of generosity' that permeated the collaborative planning process of public-private partnerships (Link, 1995). Though the culture of Silicon Valley is a popular research topic, little has been researched about the dynamics of the *communities of innovators* who work in research parks or innovation districts. Silicon Valley is a low-density, suburban model that does not provide proximity, but it does have a "pervading culture of openness and network building" (B. a. W. Katz, Julie, 2014).

220

Open source, or openness in general, communicates an inherent respect for people, and defines the community as a safe place to share ideas, dreams, and hopes with others in a collaborative capacity. In the case of Silicon Valley, the tolerance of failure and characteristic openness accumulated over decades, until it is now pervasive in the culture. The openness was an unconscious effort that flowed from the original innovators, who were seminal in the early development of the Stanford Research Park in the 1950s, 1960s, and 1970s. Now newly-emerging innovation districts recognize this cultural value of sharing and are consciously stating that they pursue and support 'open innovation' (Rainwater, 2014).

Open source was found in the communal ownership structure, the consensus based governance, and the development of social cohesiveness. The *open source* approach is conspicuous in the formation of cohousing neighborhoods and ecovillages, and then later in the execution of their governance. *Open source* is appealing and it works because it is, at its core, egalitarian and democratic.

Iterative Process

The *iterative process* is inherent in organizations focused on transformation and continuous improvement. Communitarian scholars documented this by looking across ecovillages as a whole to evaluate the function of ecovillage in relation to the community and the transformational effects on the residents (Pitzer, 1989). The evolution of *intentional communities* reflects a perpetual iteration over the decades of, not only what the term means, but also how *intentional communities* respond to a changing environment. Utopianism is —at the heart—a deep-seated belief in offering a new form of settlement that is an iteration of the current society and, while not perfect, definitely tries to be an improvement on society. *Iterative process* was evident in five characteristics of intentional communities: the development of the individual, community adaptation, new forms, utopianism, and governance.

The communities that anchor industrial districts iterate almost effortlessly out of an expertise 'in the air' that drives them as a culture to improve their products and communities (Rogerson, 1993). The research park industry recognized the changing demographics and new market pressures that forced them to embrace iteration toward a more competitive form (Townsend et al., 2009). A parallel response is seen in the emergence of the *innovation campus* concept, which is somewhat a variant of a research park. Innovation districts, due to their diversity of stakeholders and lengthy build out schedules, have flexibility and iteration built into their strategy so they can react to new opportunities as they emerge. Words have great impact, so naming a development with the prefix 'innovation' sets the expectation that inventions will be created through the *iterative process*. Disney named his community "experimental" and "prototype" with the specific intention of communicating to the residents, corporations, and the world at large that EPCOT would be the single place with a concentration of innovations not seen anywhere else in the world (S. Mannheim, 2012).

Proximity

Woody Allen, an American actor and director, is oft quoted for reinforcing the advice, "80 percent of success is showing up." It is inherently true that being in the right place at the right time leads to opportunities. Of the three themes, proximity is the distinct attribute that allows a specific university or industry to leverage the power of innovation 'in place'. The proximity theme was found in the design of cohousing, ecovillages where it was described as vital to the development of the community culture. The value was discussed at length in the literature and research on industrial districts and clusters in terms of industrial partners locating close to each other to build a network of expertise. Planners of innovation districts and innovation campuses use *proximity* as a fundamental component to network the strategic mix of assets and resources in a specified area. Evidence of *proximity* was so diffuse that the data was categorized into three types of proximity: *physical proximity*, *functional proximity*, and *emotional proximity*. Of these three, evidence indicated *emotional proximity* was the key to creating a long-standing culture of innovation.

Criticisms of the Findings

Places of innovation is a broad of a category that offers a large diversity of types of places over a 100-year time span. I reviewed industrial districts in Italy plus the United States for examples of industrial districts, clusters, research parks, innovation districts and universities. The same critical observation could be made about *intentional communities* because they took even more diverse forms in the United States and spanned 400 years. Yet diversity of data is *exactly* what gives these themes validity; the fact the

223

themes were omnipresent in various formats implies that the themes are reliable

mechanisms with wide use and wide application. The *OSipp Theory* came out of the

emerging theory process and is ready for the next phase of field-testing to develop further

research insights.

Also complicating matters is the fact that a cluster is a *construct*, whereas an intentional

community is real; it has an address and a roster of names of citizens. There is an

academic spat between the same critical observation could industrial district scholars in

Italy and cluster scholars in the United States, which is unlikely to be resolved, but same

critical observation could it does point to a very interesting difference between the

terminology and definitions. The Italian industrial districts are not a construct; each

industry is embedded in a real community with a geographic boundary and roster of

citizens. The culture element revered and celebrated in the Italian industrial district

research is just not present in the dialogue of scholars in the United States. The Italian

industrial districts are much more aligned with sustainability and holistic perspectives,

measuring outcomes beyond the one-dimensional economic metric.

Implications for Application

Implications for Practitioners

For architects, planners, and economic development strategists, this research and the *OSipp Theory* provide a wider lens to evaluate how the assets and the community can function as a whole to create a venue that is specifically dedicated to generating innovation and supporting entrepreneurship. Consider *intentional communities* as a collection of audacious utopian innovators who invest every second of their work lives and personal lives in a social experiment 24 hours per day, every day, for the purpose of developing a new prototype society. Now, consider *places of innovation* (such as research parks) as strictly work communities, organized to intentionally re-design or invent products and processes within an 8-hour day, five-day a week work setting. By blending these two ends of the spectrum, one can anticipate a different kind of model is possible; a hybrid of what is familiar in *places of innovation* and what is established as a neighborhood in *intentional communities*.

In light of the perspective provided by this research, the innovation district form is an interesting amalgam of both a *place of innovation* and an *intentional community* made of participants of the *innovation ecosystem*. An innovation district is a place of innovation that is occupied by innovators and said place typically includes mixed-use development (housing, schools, grocery stores, local businesses to meet daily needs) as well as an *innovation ecosystem* supplied to foster experimentation, dreaming, start-ups, and entrepreneurial ventures. Like the Italian scholars of innovation districts advocate, the industry and the community are entwined. Practitioners in the United States can widen their socio-economic perspective to plan accordingly.

The longevity of the development forms should also be taken into account. *Intentional communities* have a 400-year genealogy and tradition, clusters and industrial districts have a 100-year history, research parks have perfected their development recipe for 50 years and innovation campuses have existed for just two or three decades. Innovation districts in the United States are less than ten years old and the vast majority of these initiatives are still in various stages of exploration, formation, and planning. As the leading voice of innovation district documentation explains, "To date, networks of innovation district practitioners and leaders remain nascent and isolated" (B. a. W. Katz, Julie, 2014). Because they are embedded into a community, urban innovation district participants are much more diverse than a typical research park. The pace of work varies for each collaborator in an innovation district: by semester for the academic partner, by work day for the city administration that works from 9 am to 5 pm, Monday through Friday, year round, and by no calendar, weekday, or day time constraints for the entrepreneurs driven by a 24/7/365 fervor. The community overlay of innovation ecosystem services has many masters to serve. Any coordinated effort to design such a community requires masterful skill on the practitioner's part. Appreciating the genealogy of the innovation district and the unique mix of participants is a contribution this research makes. The *OSipp Theory* introduces *open source*, *iterative process*, and *proximity* as design goals that can be incorporated by the practitioner into the strategy, operation, and management of an innovation district.

Example of the *OSipp Theory*.

Emerging innovation districts often use the strategy of deeply appreciating and understanding their strategic partnerships by exploring their capacities fully, defining the roles, and delegating responsibilities. This management of expertise is the mark of solid collaboration. They also seek specific opportunities to achieve demonstrative "early wins" so the greater community can see what an innovation district contributes. As an example, McGill University has entered into agreement with the Province of Quebec, the City of Montréal and ETS (L'École de Technologie Supérieure) to establish an Innovation District (Quartier de L'Innovation) in the Griffintown area of the city. As part of this agreement, McGill set out to be clear as to their own specific role and how it differed from the other participants. The University viewed the biggest opportunity as coming from the rich and diverse collective experience and perspectives that the participating partners bring to the QI initiative and, at the same time, saw the biggest challenge as deciding how to effectively and productively use the different skills of these organizations in a timely and strategic manner so that the QI would become a reality.

The outcome of this analysis led McGill to choose roles and responsibilities that were best suited for the University and that would help the QI come to fruition. More specifically, McGill choose to focus its resources and efforts on creating an innovative district by building upon the school's values, skills, diversity, and experience. This also meant incorporating McGill's long view of the sentiment expressed in their core commitments: Ideas, Innovation, Sustainability, Collaboration and Partnership & Social

Engagement ("Quartier de l'innovation: a joint vision for a prosperous future," 2014).

At the core of their approach to the innovation district, McGill committed to the idea of *open source,* where "research ultimately leads us, and all advancements begin with ideas" (Brabant, 2012). In an email exchange shared between the University and the consultants it was explained that this commitment was founded in the belief that the "pursuit of knowledge and fundamental discovery helps us to better understand the world and take incremental steps, sometimes in unexpected ways, toward a better future. As such, we remain steadfast in our commitment to curiosity-driven research" (Brabant, 2012). In doing so, McGill set out the goal of placing greater emphasis on innovation in all its forms—social, pedagogical, and organizational—as well as the development of new technologies. They did so by committing to facilitating, encouraging, supporting, and rewarding research partnerships across academic fields, both on the campuses and with external partners.

McGill chose to focus on the specific social needs of the existing communities within the district and in conjunction with the learning opportunities of the University faculty and students. They held the intention "of endeavoring to apply research to shared challenges; providing innovative learning environments for students at all levels; and seeking out and supporting initiatives that result in tangible improvements for individual communities" (Brabant, 2012). This represents the opportunities that arise through *proximity.*

McGill focused on developing specific pilot projects that would 'innovate in place'.

228

The "volet urbain" —or urban aspect of the QI initiative—was considered to take many forms, both in subject matter and in type of project. Focusing on a small number of projects was necessary to maximize effectiveness. The projects were selected for their potential for innovation, their relevance to urban sustainability, their potential for interdisciplinary collaboration, and their links to ongoing activities at McGill. All projects would pertain to both physical development (the actual building of the QI) and research about that development that advanced the knowledge and practices of innovation district operations.

Most of these innovative pilot projects focused on the relationships that provided learning through the *iterative process*. They were seen as an ongoing laboratory to explore high quality, clearly defined, indicators for operationalizing, measuring, and assessing progress toward sustainability and innovation, which are often ill-defined concepts in cities. For instance, Griffintown benefits from the presence of district heating, *proximity* to downtown and the Metro light rail transit, and availability of land for new, exemplary development. In turn, urban energy use is a key topic of interest in academia, industry and government. The *functional proximity* results in innovations in energy use because the university provides research for the community needs. With performance measures becoming increasingly important, both practically and politically, an opportunity was seized to develop a set of indicators for the Quartier de L'Innovation that would benefit the QI project itself through monitoring, while advancing general knowledge about district heating (Fischler, 2013).

Starting from scratch.

After considering who the participating partners are in an innovation district, another question for the practitioner turns to understanding what the innovation potential is as it correlates to an educated public. According to Edward Glaeser, an economics professor at Harvard who is often quoted in innovation district publications, "it is very hard to imagine how you can have anything that can be plausibly called an innovation district if 10 percent of your adults have college degrees. It's all about having smart people who are connected by urban density and who learn from each other and work with each other " (Pazzanese, 2014). For the practitioner, this presents the unresolved issue of defining what an innovation district is, and what it can be. Shearmur argues there are fast innovators and slow innovators; the former can only operate in a dense urban environment, whereas the latter can operate in urban or rural environments (Shearmur & Doloreux, 2016). It then becomes a matter of being able to develop a strategy to develop an *innovation ecosystem* based on asset mapping to be able to access feasibility. Although the first wave of innovation districts has occurred in larger, dense cities that does not necessarily imply that smaller cities or even rural areas cannot develop their own interpretation of an innovation district. Case in point, Chattanooga, Tennessee—population of 176,000—has positioned themselves as a model for the mid-size city innovation district (Glenn, 2016). Mayor Andy Berke is a frequent speaker on the national platforms for innovation district planning.

The preplanning phase and the social programming strategy are where the *OSipp Theory* (*open source*, *iterative process*, and *proximity*) may prove to be fundamental to establish the initial culture of the innovation district. By applying the *OSipp Theory* to this newly emerging form, loosely referred to as innovation districts, we can postulate how these *communities of innovators* would manifest. Established innovation districts can use the *OSipp Theory* as a cultural benchmark to evaluate the functional health of their innovation ecosystem within the community.

The main message the *OSipp Theory* sends is a platform of trust. It attracts minds that think alike and share core values (*emotional proximity).* It provides programming that networks them (*functional proximity*). It establishes that participation is valued and expected (*open source*), and most importantly, it creates a culture that is tolerant of innovative efforts based on cycles of improvements on "failures" (*iterative process*). Specifically, the mechanisms might unfold as *open source* information exchange being organized around the sharing of intellectual capital and economic resources in pursuit of improving the greater *innovation ecosystem*. The community and participants in the *innovation ecosystem* would also continuously collaborate to improve the business environment within the public realm of the innovation district, which, in turn, supports economic development and quality of life. The entrepreneurial network would use *open source* practices to support each other and improve the fertility of the innovation district.

Iterative process could be used in the capacity building stages of preplanning and later in the methodical execution. The inclusive nature of the *open source* platform would likely generate an ever-expanding circle of inspiration and invitation. The job of inclusion and continuously incorporating learning curves is never finished. The guiding strategic documents of the development strategy might reflect a refreshing flexibility as new resources become available and as the external environment changes. It would, from the outset, allow for the constant ebb and flow of resources from the participants and the community in much the same fashion as high-turnover *intentional communities* do; many have come to accept it, celebrate it, and leverage the fresh perspectives turnover provides (Aguilar, 2012).

Proximity plays a vital role in an innovation district as mass transportation creates a circulation of social capital within the geographic bounds. This same delineated area concentrates the resources needed by the innovators in the *innovation ecosystem,* and it also increases the likelihood of knowledge spillovers. This critical mass is necessary to "choreograph collisions" as the language of planners implies. The boundary of an innovation district maintains a permeable edge so there is potential to spark adjacent areas of the city or serve as inspiration for a separate pocket of development. When transportation routes define the geography of an innovation district, they do so by providing a circular radius that invites all who care to walk rather than by imposing a strict boundary line on a street that defines one side of the street as an innovation district and the other side as not. Table 7 provides a summary of how the *OSipp Theory* could be applied strategically to some future next evolution of an innovation district focused on innovations around sustainability—of a *sustainability-oriented innovation* (SOI) district.

Table 7. The OSipp Theory Potential Forms in an SOI Innovation District

- Open source: Organized around sharing of intellectual capital
- Open source: Community and public support for economic development
- Open source: Ecosystem integration with resources

- Iteration: Inclusive capacity building involving all stakeholders
- Iteration: Living documents of development strategies, benchmarks, and goals
- Iteration: Constant ebb and flow of resources from city, college, corporations

- Proximity: Social programming to circulate people, ideas, and resources
- Proximity: Geographically delineated by public transportation routes
- Proximity: Permeable edge with potential to spark other pockets of city

Implications for Higher Education

Higher education has spent a few decades allowing handfuls of passionate professors to incorporate sustainability into their curriculum and many of those sustainability advocates (professors and staffers) have created minors, certificates, and degrees to appease the students for whom the one token sustainability class was not sufficient. Universities have incorporated sustainability principles into the facilities management side of campus operations where said practices proved to reduce waste and save money. Most universities approach the curricular development of sustainability incrementally with token efforts, but rarely at a strategic, institutional level. Very few have recognized or embraced their leadership role of being a national force that is responsible for educating for sustainability, much less for fostering innovations specifically around sustainability (Cortese, 2003). The moral imperative of sustainability does not exist in the general public so why would it be prevalent in higher education? The definition of a leader is one who is ahead of the others and has the foresight to know which direction to head. Can higher education institutions be the leaders or co-leaders in advancing

sustainability?

To see the opportunities, a university has to view itself as part of a broader community

with larger responsibilities than just their two primary missions of education and

research. Service, its third mission, is increasingly being expected to be fulfilled by a

concerted effort to help stimulate and to sustain economic development through regional

innovation (Gibson, Foss, & Hodgson, 2014). Place-based knowledge is not easily or

quickly replicated and thus presents a competitive advantage for a savvy university

(Amin, 2004). Though this seems obvious, too often it goes under-appreciated and under-

utilized. Also, universities are finding themselves cast into the role of urban planners;

they develop adjacent neighborhoods that "provide jobs, housing, services, and

entertainment for residents, many of whom have no academic connections" (Campbell,

2005). For two hundred years, the typical campus in the United States was built to be

self-contained and closed off from the surrounding community; the design of University

of Virginia built in 1817 became the model for this approach (Puddu & Zuddas, 2013).

With their physical form, they "turned their back on the cities that surrounded them in

order to isolate the academic community" thus ensuring uninterrupted concentration

(Way, 2016). Clearly, to become a 'city of knowledge' in the twenty-first century, those

physical borders will need to become more porous and more inclusive.

A university can serve as a development catalyst, but that requires institutional flexibility.

Traditionally, universities see themselves as "holders, creators, and diffusers of knowledge—not as mobilizers and catalyzers of knowledge" (Allison & Eversole, 2008). Already, the co-creating model of knowledge generation and production development is apparent, as is the recognition that place-based knowledge hubs have the potential to provide unparalleled educational experiences *because* of their place (Allison & Eversole, 2008).

The university could re-imagine their housing as an opportunity to provide a venue for experiential education for sustainable living or a venue to foster *sustainability-oriented innovation* and research. Fostering discontinuous or radical innovation means creating a space and a social contract where *iterative failure*, as with rapid prototyping failure, is expected and accepted. Kanter refers to a classic industrial innovation study that showed that up to 3,000 raw ideas could be required to produce even one successful new product (Kanter, 2012). As a 'living learning community' advocate at Virginia Tech observed, "There are few environments where students will spend more time than in a resident hall. The potential impact on student learning is enormous" (Shushok Jr et al., 2013). In 1904, the Harvard student affairs officer quoted Harvard's president regarding the decision facing university housing as integral to the educational mission:

> We are come to the parting of the ways, where we must either make up our minds that the social life of the students is none of our affair—and in that case we had probably better give up the college as an institution altogether, and confine ourselves to the work of the schools which prepare men for practical life; or we must bring our men together into a real community, with a common life – a true college life. (Lowell, 1904)

The university could take into account the wealth of intellectual capital that is eager to return to the campus in the form of retiring alumni. They could evaluate the presence of tacit knowledge embedded within the social networks of the campus community and

235

regional culture. They could consider unique collaboration potential of co-location that allows industry partners to physically become part of the campus community. As the university shifts their business model to serving lifelong education consumers, they could begin to embrace the concept of being the anchor of a continuous and evolving community. This community could also model utopian ideas of living sustainable lifestyles and actively creating *sustainability-oriented innovations*.

This bold vision could take the form of an innovation district on campus, an innovation district adjacent to campus, an innovation campus embedded in a complete residential mixed-use community, an academic ecovillage, or something entirely new—such as an innovation village or an 'experimental prototype community of tomorrow'. Given the sizable investment required to create a physical place for a critical mass to converge, there is not a large margin for error in the iteration of the kind of large prototype community that would lead to a full-scale functioning model. Whichever course of action is selected, it will likely be an audacious project with no precedents to emulate. That action will be audacious for its novelty and complexity, but yet that is what is predictably next as an incremental innovation. An audacious action is the logical progression based on the trajectories already established. Figure 7 illustrates how large the impact can become from the intersection of sustainability, innovation, and community.

Figure 7. Full Diffusion of SOI from within Community

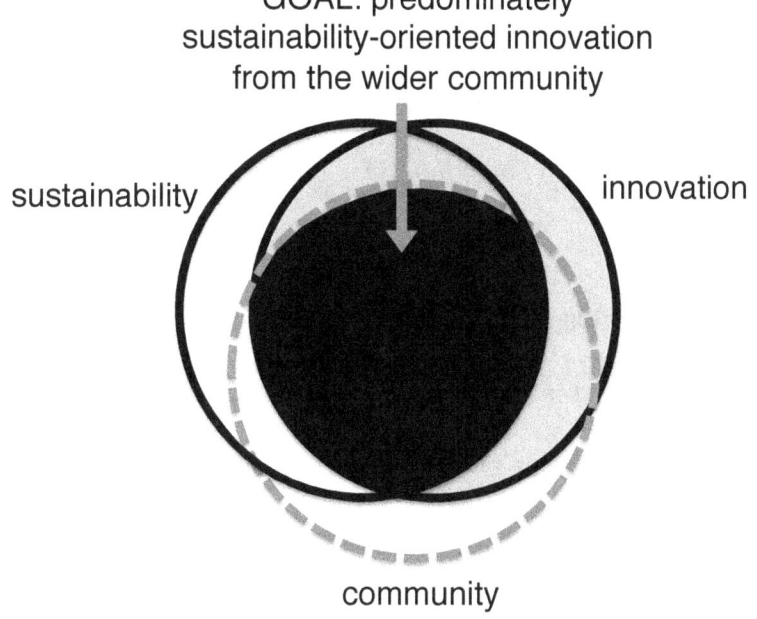

Implications for the academic ecovillage.

Most ecovillages do not pursue innovation on the cutting edge of technology, but rather

they find value in integrating the social, technological, and governance into one system.

As veteran ecovillage traveler Daniel Greenberg observed about blending low tech and

high tech at ecovillages, "What makes ecovillages unique and relevant is how they are

putting these pieces together into wholes that are greater than the sum of their parts" (D.

Greenberg, 2013a). This holistic attempt positions the ecovillage as a 'campus' where

students can travel to visit for an immersive experience. In 1999, the nonprofit

organization Living Routes was founded to provide study-abroad experiences that

qualified for course credit through the University of Massachusetts-Amherst.

Reinterpreting an ecovillage on private property as an 'academic ecovillage' on a

university campus—as a choice among the residential housing venues the university provides—would be just one way a university could increase practical understanding of sustainability for students. For an excellent assessment of the symbiotic relationship potential from pairing weaknesses with strengths of academia and ecovillage, review Daniel Greenberg's book chapter *Academia's Hidden Curriculum and Ecovillage as campuses for Sustainability Education* (D. Greenberg, 2013a).

Privately-funded ecovillages struggle to survive past their formation stage due to regulatory, financial, and interpersonal hurdles that conventional neighborhood developers do not face; in fact, 90% of the ecovillages fail (Christian, 2003). The privately-funded ecovillages that survive defy conventions of the urban development mainstream by excelling at alternative construction, social governance, and resource management; three skill sets useful in the resource-constrained future (Boyer, 2015). A university-funded academic ecovillage would likely not face the same regulatory hurdles, experience the financial strain of securing investment, or have the problems generated by group decision making.

In 1962, prior to the emergence of the back-to-the-land communes of the late 1960s and the ecovillages of the 1970s, there was a call for the United Nations to invent and fund "scientific intentional communities" to serve as models that would be "instructional by 2025" (Winthrop, 1962). Walt Disney's lifelong dream of designing and building a city for innovators—the Experimental Prototype Community of Tomorrow (EPCOT)—was never realized due to his death in 1966 (Chytry, 2012). Place-based innovation became the domain of university research parks in the latter half of the twentieth century, but these places were office parks not communities of citizens. Around the turn of the twenty-first century, a new scholarly field emerged from within the management literature: eco-innovation and green innovation. These two lines evolved into *sustainability-oriented innovation* topics and emerged as a subtype of innovation that focuses energies specifically on inventions that solve not only environmental problems but also address social equity and economic goals (Jay & Gerard, 2015). Perhaps the merging of all of these concepts—the scientific intentional community, the EPCOT city of innovators, place-based innovation, and *sustainability-oriented innovation*—can lead to the creation of an innovation village model for the broader university community of town and gown.

If a dream is considered unachievable, or if every experiment is expected to fail, that may explain why sustainability is an intriguing topic to discuss on university campuses, yet an elusive one to pursue in practice. History has left us nuggets of wisdom about the role of expectations. In the great gyre of information that now exists, that wisdom is still accessible to those with discernment to recognize it and vision to implement it. A popular quote attributed to Henry Ford is "Whether you believe you can do a thing or not, you are right" —which is more succinct than his actual original quote. In 1925, Henry Ford said, "You must never, even for a second, let yourself think that you can fail. Our first principal is that failure is impossible. You may not get what you're trying to do right the first time or the second time or the tenth time or the 100th time, but if you shut out of your mind the possibility of being licked, then you are bound to win" ("Stick to the finish," 1925).

If higher education, as a whole, believes they will not succeed in any attempt to model sustainability at the community level, they will fail. If higher education does not even attempt this experiment, they will fail. Therefore, to avoid guaranteed failure, higher education must try because the only failure is the failure to try. But the industry-wide response from higher education does not need to manifest in unison, it needs to begin with just *one* institution to execute an audacious experiment, with Disney-like imagination and the Ford-like conviction, that the impossible is realized through the appropriate mindset and commitment to iterations. The *OSipp Theory* provides an appropriate approach to foster innovation and the Earth Charter provides the necessary values to anchor and guide actions.

Implications for sustainability faculty.

The beauty of the *OSipp Theory* (*open source, iterative process, proximity*) is its ability to potentially scale for a smaller application such as a classroom experiment, instructor pedagogy, or a departmental strategy. When the culture of the *OSipp Theory* permeates the university, everyone can participate: the faculty, the staff, the administration, the community, the alumni, and the students. They literally have the tool to invent how to apply the principles. The idea of progress, the ability to engage in 'social dreaming', and the quest of perfectibility becomes beneficial in many settings. Pedagogy in the classroom employing the *OSipp Theory* becomes the platform for the co-creation between the faculty and the students that creates a novel learning experience (Prahalad & Ramaswamy, 2004). Under the *OSipp Theory,* mistakes and failing scores are not failures but beginning points that become platforms for iteration until true comprehension and

240

learning is achieved. Networks of trust are formed, not between students just sitting in *physical proximity* to each other, but though genuine collaboration that creates *emotional proximity* and *functional proximity*.

All the future generations, not just the current one, require education in sustainability. They are going to have to think of new concepts that are not explained in the textbooks. How do educators train them to approach such a nebulous task? The *OSipp Theory* is an approach to help spur innovative thinking. Albert Einstein left a sobering thought for educators,

> It is not so very important for a person to learn facts. For that he does not really need a college. He can learn them from books. The value of an education in a liberal arts college is not the learning of many facts but the training of the mind to think something that cannot be learned from textbooks. (Frank, 1951)

Implications for corporate co-location on campus.

The advent of corporate co-location—where a corporate partner shares a building with academics and students—has co-evolutionary implications. Co-location is defined as the "intentional co-location of academics and industry to facilitate and streamline the commercialization process" (Bramwell et al., 2012). All parties *will* change from the experience of forming a new community of innovators. Norgaard explains the co-evolutionary paradigm as, "In biology, co-evolution refers to an evolutionary process based on reciprocal responses between two interactive species … the concept can be broadened to encompass any ongoing feedback process between two evolving systems … Co-evolution occurs when at least one feedback is changed, which then initiates a reciprocal process of change" (Norgaard, 1984).

241

In an ecological system, there is no way to predict how these changes will manifest (Norgaard, 1984). In a community of innovators, the evolution can be directed by agreed-upon guiding principles, as seen in individual experimental communities, or even the founding of a new republic as seen in the United States in 1776. Whether guiding principles are borrowed from an existing framework (such as the Earth Charter or the One Planet Living initiative) or created by collaborative agreement, it is vital that a set of principles be articulated and agreed upon, produced in writing, and implemented consistently until they are an inseparable part of the culture.

The goal of corporate co-location is to develop a synergistic relationship where all parties benefit by the *physical proximity* and association. A win-win situation is possible, probable, and even preferable. The axiom "iron sharpens iron" does not happen by accident, but rather is the result of two strong entities that agree to continuously improve through iterative exchanges. Choosing this co-location partner requires a deliberate, systems-thinking analysis that is capable of looking at the long term developments and thinking through the possible consequences. Scenario planning is a useful planning tool to forecast the possible outcomes of corporate co-location. Also, the *OSipp Theory* can be utilized to ensure a mutually agreed upon method of generating innovation. The *OSipp Theory* initiates a dialogue based on transparency due to the *open source* strategy. This platform allows ideas to percolate, concerns to be heard, precautions to be taken, and even warnings to emerge from the stakeholders of corporate co-location.

The exploration of a potential corporate co-location partner should be an *iterative process* itself. Collaborations should produce versions of the possible futures, but allow for time in between the iterations for the ideas to settle and re-emerge slightly improved. Gradually, the desired optimum relationship will congeal or the sub-optimum relationship will disintegrate, which is best for all parties concerned in the long run. And lastly, *proximity* should be utilized frequently. Working through a co-location selection and agreeing upon goals and methods requires in-person dialogue. Co-location between a university and a corporation is as serious – if not more serious – than co-habitation prior to marriage between two people. Before 'shacking up together', the university stakeholders and corporations should spend time on smaller initiatives to evaluate if their stated philosophies of business, cultures, goals, and methods align in practice.

To continue to the biological metaphor, co-evolution is an *iterative process* that evolves over time; it cannot be rushed. It is a response to resources. Darwin's *Survival of the Fittest* Theory refers to those species that are most appropriately matched to the environment and are able to reproduce. The question then becomes, what does the co-location relationship want to create? Is it world-changing students, commercialization, revenue, social networks, employee vetting, grant potential, advanced knowledge? The goals and the methods are a subject for agreement in advance.

Recommendations for Future Research

The first time a field scientist makes a site visit to a new ecosystem, it is unknown what exact observations will be noticed and recorded. Even with piles of books and years of preparation, even seasoned researchers are surprised by what they observe when physically embedding themselves in a place. Often the instrument that emerges is a field note journal in chronological diary format. An explorer has a different experience. There is more of an open-ended agenda. A researcher who explores pursues an inductive investigation.

As discussed earlier in the limitations section of chapter 1, content from dozens of valuable interviews about visioning futures was not captured due to not recognizing the research opportunity presented during an academic ecovillage planning project hosted by the university ("Wake Up & Dream project to host advocate of ecovillages across the world," 2012). Only after the conclusion of the 12-month exercise was it obvious that tremendously robust data were generated. On the front end, the entire project was an absolute unknown in terms of participation as well as quality and quantity of data. It was the first high-profile exploration of its kind in higher education. To the next institution that undertakes such an audacious exploratory process around ecological housing or sustainable community design, I would recommend allowing adequate time to formulate a formal research agenda to run in parallel to the planning explorations. Though capturing statements may hamper creative input and unedited participation, the value of recording the insights of participants and thought leaders would contribute substantially to the under-researched field of sustainable communities involving higher education institutions.

Having spent over a decade observing sustainability in practice as well as reading about it in hundreds of books and articles, I have come to the conclusion that the only place sustainability *can exist* is in practice. Sustainability is like a fish; it can be studied as a specimen in a book or dried and mounted, but can only live in water and is only truly understood if observed in water in the full context of its native habitat. At this point of revelation, armed with field observations, I left the 'field' of practitioners and pursued exploratory research of what was to become the *OSipp Theory*. My intent was to describe what I intuited from the practitioner's realm and to substantiate those observations with stories that reflected those themes found in academic literature. To join the theory and the practice strengthens both, and contributes to the advancement of sustainability.

The *OSipp Theory* research generates *more* questions. Do these innovation communities recognize they are using the *OSipp Theory*? Do these innovation communities ever fail using the *OSipp Theory* approach? Can the *OSipp Theory* be applied to achieve other purposes besides innovation or *sustainability-oriented innovation*? The *OSipp Theory* is a problem-solving process with potential for multiple applications, though it is likely unrecognized by organizations that have used it, or partially used it, because *open source, iterative process*, and *proximity* can—and does—come from the inherent nature of some cultures. The *OSipp Theory* could be applied to nefarious goals as well, because it is an effective approach, regardless of the intention the organization might have. For this reason, it is suggested the most appropriate use of *OSipp Theory* should be rooted in the sustainable values of the Earth Charter so that it can spawn positive progress toward sustainability.

Values are key to provide direction for the future waves of innovations. Without values, innovations are merely profitable novelties that advance the status quo of business that is unsustainable and ultimately destructive to humanity. With values, the innovations of the future can be intentional and goal-oriented to produce *truly* sustainability solutions. David Orr lists values first: "the kind of education we need begins with the recognition that the crisis of global ecology is first and foremost a crisis of values, ideas, perspectives, and knowledge, which makes it a crisis *of* education, not one *in* education" (David W Orr, 2002).

My desire is that the three themes identified will be applied by future researchers as an emerging theory worthy of deductive testing through interviews with visionaries who plan innovation cultures for cities, universities, classrooms, and corporations. The *OSipp Theory,* presented as a set of values, can have enormous implications when they are articulated on the front end of a planning project. A value, once declared publicly, has the power to guide a vast array of investment decisions; at that point all efforts become guided by those stated values. Using the *OSipp Theory* as a North Star guide, it becomes a simple matter of evaluating each potential action as moving a university toward or away from the specified goal. If the *OSipp Theory* is already being applied consciously by practitioners, then there is an opportunity for field research to validate its impact through direct, quantifiable research methods.

At the very early stages of this the doctoral journey, the patterns of *open source*, *iteration*, and networked *proximity* were originally observed in practice in Vancouver, British Columbia, a city that shifted itself from relative obscurity in the 1970s to become world-renowned for sustainability by the year 2000 (Scerri & Holden, 2014; Walsh, 2013). Further research found these same three themes to be prevalent in *places of innovation* and in social movements in sustainability as certain industries pioneered those early efforts in the built environment, corporate social responsibility, and higher education. To investigate the strategic premise that these three underlying themes are inherent in innovation in place, I designed and implemented a one-year planning exercise amongst stakeholders at a university to explore how sustainability could be taught and practiced in the context of a specific place; in this case focusing on the aspiration of a university-owned academic ecovillage. The investigation in this dissertation confirmed that *open source*, *iterative process*, and *proximity* are inherently present in a variety of sustainability initiatives that were analyzed: planning practitioners, my academic ecovillage planning exercise, and industries advocating sustainability. But still, what was evident in practice needed to be anchored by theoretical evidence articulated by research. Thus, what emerged was the framework of this dissertation.

Future research could explore the *OSipp Theory* on a sample set of innovation districts by developing a metric to measure the degree to which the each of the themes is present at different stages in the development of an innovation district. Rather than establishing a benchmark and a level of completion, a tool could be developed that allows a metric that tracks iterations and development. As Disney says, a community of innovators is always in a state of 'becoming'.

Conclusion

The intent of this research is to understand more clearly how others in the past approached what was perceived as unimaginable, so that this generation can build the confidence and the courage to tackle humanity's sustainability goals that now seem unachievable. History provides the perspective to look at human development as a series of innovations—product innovations, technological innovations, medical innovations, social innovations, and civic innovations—which were the result of cumulative knowledge and collaborative efforts by groups of people, recognized in this research as '*communities of innovators*'.

At issue, though, is the fact that most products and processes used in modern civilization were designed and developed without sustainability principles as a parameter. If this paradigm is perpetuated, the inescapable result will be an exhausting of the planetary resources and compromising of the Earth's ability to provide a habitat for the human race. Sustainability is *the* key challenge facing humanity in the twenty-first century because of ecological instability, social equity unrest, and economic systems that operate beyond the boundaries of sustainability principles. Sustainability requires managing finite resources and accounting for the impact associated with externalizing or delaying the true costs of commerce.

As science advances to provide a clearer understanding of our ecological standing and mankind's consciousness evolves toward responsible stewardship of Earth, there exists a potential for this generation to "essentially, completely change the world" (Z. Goldsmith,

2007). The global society now finds itself on the cusp of a revolution of innovation, literally, before them awaits a massive need for sustainability-oriented innovations because that is what is necessary to restore an ecological equilibrium within the economic system.

For a better understanding of how to foster this innovation revolution, the research began by looking at recent forms of *communities of innovators* in the built environment. In the *intentional community* realm, the research began with ecovillages and then traced their genealogy back through the *intentional community* movement over the course of history in the United States. Among *places of innovation*, the research started with innovation districts and then an understanding was developed of previous venues for innovation over the past century.

The research question driving the study is: "What are the mechanisms historically used by *communities of innovators* as identified in *intentional communities* and in *places of innovation* that were used to approach their goals?" The inquiry was then tailored for the university setting. In the university context, the question driving the investigation is: "How can these mechanisms be applied to the environments created by higher education institutions so they can successfully fuel innovations that advance sustainability?"

A qualitative methodology was applied to conduct a historical analysis of past *communities of innovators*. The data was analyzed using the process tracing method borrowed from the field of Political Science. The data was gathered under the approach of historical methodology. In all, over 1,200 sources of evidence were reviewed over the course of iterations between the working hypothesis and the data collection.

The key findings led to the recognition of three themes consistently present in the two places known for having an experimental nature: *intentional communities* and *places of innovation*. In all, eleven types of *communities of innovators* were considered: intentional communities (also known as experimental communities), communes, ecovillages, academic ecovillages, cohousing, innovation community prototype, industrial districts, clusters, research parks, innovation districts, and universities. The three themes that emerged were: *open source*, *iterative process*, and *proximity*. Combined into one approach, this created the *OSipp Theory* for fostering innovation. A working definition of the *OSipp Theory* is an approach intended to foster innovation by combining mutually reinforcing mechanisms of *open source* for information exchange used with an *iterative process* to finding solutions, which are utilized by a community of people who interact through a physical, functional, or emotional *proximity*.

For those people tasked with fostering innovation 'in place', this research generates implications about how to apply the *OSipp Theory* in the attempt to catalyze *sustainability-oriented innovation*. Higher education institutions are in a unique position to respond to sustainability imperatives (Cortese, 2003). In the current dynamic situation —of disruptive competition and the demands of the *knowledge economy*—higher education institutions are faced with complex choices about how to secure the relevance of their institutions. Embracing the leadership role of advocating *sustainability-oriented innovations* is one potential strategy that serves the interests of many.

It is said that no one likes, yet people embrace change when it is presented as innovation. Buckminster Fuller says, "You never change things by fighting the existing reality. To change something, build a new model that makes the existing model obsolete." Sustainability is comprised of decisions that are intentional acts of wisdom; it remains to be seen if the global society is wise enough to put energies into innovations that address issues and promote sustainability, rather than create more issues.

Critical Evaluation of the Research

To develop an emerging theory required, for me at least, becoming comfortable with excess amounts of ambiguity over a long period of time and through many disciplines and many industry experiences. Like mining for diamonds, it requires abundant searching and, in the end, you still only have a diamond in the rough. This emerging theory is still just an untested prototype that now requires refinement and field-testing, specifically in innovation district developments and specific innovation projects on campus. The *OSipp Theory* can be improved, refined, and even come with written operation instructions. This kind of multidisciplinary research is better suited for a collaborative team, whose every member is already deeply versed in a topic, rather than a PhD dissertation by a solitary author.

How this Study Changed the Researcher

I have a new appreciation for how the field of history developed over the past century and how the historical methodology can be universally applied. I now consider historical methodology to be the *single most useful skill* I learned during my graduate experience at the university; I wish I had been introduced to it earlier in my academic career thirty years ago.

I also greatly appreciated the scholars who voiced contrarian views. It is very easy to agree with the majority and to craft research that reinforces the dominant paradigm. But the research that voices humility or admits the unknown or conjures up original ideas, earns my attention and respect. It is my own anecdotal observation that the older literature (pre 1970s) is written less to impress and more in the scientific spirit of discovering incremental truths. Interestingly, contributing incremental truths is the basis of the theoretical framework of scientific realism, which is the underpinning for the process tracing method I used in historical methodology.

Overview of the Importance

The values articulated in a strategy are foundational, but all too often overlooked for more tangible practical steps, yet they are key to achieving goals. Values keep the future actions aligned to initial intentions and this develops a depth of character and long-term consistency to a community. Higher education—more than ever before—is in a unique position of leverage in setting the course of history.

Consequences of not developing this field of research.

Without a learning curve or insights gleaned from research, much energy can be wasted prototyping various strategies designed to increase innovations around sustainability. The themes offered in this dissertation, crafted in the *OSipp Theory*, are the CliffsNotes, a few short cuts, and a springboard for those in the years to follow who are keenly intent on fostering *sustainability-oriented innovation*.

Leo Tolstoy (1828-1910) understood that the obvious, just isn't. Tolstoy, speaking about the interpretation of art, also makes a lucid observation that offers a timeless perspective:

> I know that most men—not only those considered clever, but even those who are very clever and capable of understanding most difficult scientific, mathematical, or philosophic, problems—can seldom discern even the simplest and most obvious truth if it be such as obliges them to admit the falsity of conclusions they have formed, perhaps with much difficulty—conclusions of which they are proud, which they have taught to others, and on which they have built their lives. (Tolstoy, Pevear, & Volokhonsky, 1995)

This research attempts to inform higher education institutions about sustainability strategies—the vital importance of it, how to foster innovations that leads to it, and whose responsibility it is to educate for it. It is a topic that needs to be fully developed so it attracts more scholars who can diffuse the ideas and carry the wisdom into leadership positions as they advance in higher education. The courageous leaders in higher education need to be given validation for forward-thinking ideas and for making a chorus out of singular voices.

Who benefits from this research.

Future generations of students, campus faculty and staff, regional economies, firms, local communities—all have a role to play in the collaborative efforts and they have benefits to reap from participation. Ultimately, if these efforts are successful, and higher education actually does generate the necessary innovations to create a sustainable society, then the 9 billion people on the planet in 2050 will prosper based on the intentions that are incorporated into higher education today. It all starts with an intentional effort based in noble utopian aspirations. The *OSipp Theory* reintroduces those utopian aspirations.

Recommendation – Key Message

The first step toward realizing sustainability-oriented innovations 'in place' is to understand the specific mechanisms and opportunities that arise from strategic and complimentary partnerships, in conjunction with the *proximity* of assets and attributes. The *OSipp Theory* frames the mechanisms that bridge the stated values and the stated goals of a *community of innovators*. The fundamental values of *open source*, *iterative process*, and *proximity*, central to applied sustainability and innovation, assure that place matters, intention is direction, and relationships are at the core of collaboration.

REFERENCES

Abramovsky, L., & Simpson, H. (2011). Geographic proximity and firm–university innovation linkages: evidence from Great Britain. *Journal of economic geography*, lbq052.

Abrams, P., & McCulloch, A. (1976). *Communes, sociology and society*: CUP Archive.

Adams, R., Jeanrenaud, S., Bessant, J., Denyer, D., & Overy, P. (2015). Sustainability‐oriented innovation: a systematic review. *International Journal of Management Reviews*.

Aguilar, J. (2012). Assessing Success in High-Turnover Communities: Communes as Temporary Sites of Learning and Transmission of Values. *Journal for the Study of Radicalism, 6*(1), 35-57.

Ahn, Y. (2005). Research Parks and Economic Development. *Urban and Regional Economic Development Handbook.* Retrieved from http://www.umich.edu/~econdev/

Aiken, G. (2012). Community transitions to low carbon futures in the Transition Towns Network (TTN). *Geography Compass, 6*(2), 89-99.

Aldrich, R. (2012). Wisdom Way Solar Village: Design, Construction, and Analysis of a Low-Energy Community: Golden, CO: National Renewable Energy Laboratory.

Alexander, A. (2015). The cohousing association of the United States. Retrieved from http://www.cohousing.org/dırectory

Alexander, C., Ishikawa, S., Silverstein, M., Jacobson, M., Fiksdahl-King, I., & Angel, S. (1977). A Pattern Language: Towns, Buildings, Construction (Center for Environmental Structure).

Allen, J., Nelson, M., & Alling, A. (2003). The legacy of Biosphere 2 for the study of biospherics and closed ecological systems. *Advances in Space Research, 31*(7), 1629-1639.

Allen-Gil, S., Walker, L., Thomas, G., Shevory, T., & Shapiro, E. (2005). Forming a community partnership to enhance education in sustainability. *International Journal of Sustainability in Higher Education, 6*(4), 392-402.

Allison, J., & Eversole, R. (2008). A new direction for regional university campuses: catalyzing innovation in place. *Innovation: The European Journal of Social Science Research, 21*(2), 95-109.

Amin, A. (2004). An institutionalist perspective on regional economic development. *Reading economic geography*, 48-58.

Andreas, M., & Wagner, F. (2012). "For Whom? For the Future!" Ecovillage Sieben Linden as a Model and Research Project. *Realizing Utopia*, 135.

Archer, L. B. (1964). *Systematic method for designers*: Council of Industrial Design.

Audretsch, D. B., & Feldman, M. P. (1996). R&D spillovers and the geography of innovation and production. *The American economic review, 86*(3), 630-640.

Audretsch, D. B., & Keilbach, M. (2007). The Theory of Knowledge Spillover Entrepreneurship. *Journal of Management Studies, 44*(7), 1242-1254.

AURP. (2016). Association of University Research Parks. Retrieved from http://www.aurp.net/

Bader, C. D., Mencken, F. C., & Parker, J. (2006). Where Have All the Communes Gone? Religion's Effect on the Survival of Communes. *Journal for the Scientific Study of Religion, 45*(1), 73-85. doi:10.2307/3590618

Bagnasco, A. (1977). Tre italie.

Baker, T. (2013). Ecovillages and Capitalism: Creating Sustainable Communities. *Environmental Anthropology Engaging Ecotopia: Bioregionalism, Permaculture, and Ecovillages, 17*, 285.

Baldwin, C., & von Hippel, E. (2011). Modeling a paradigm shift: From producer innovation to user and open collaborative innovation. *Organization Science, 22*(6), 1399-1417.

Baptista, R., & Swann, P. (1998). Do firms in clusters innovate more? *Research policy, 27*(5), 525-540.

Barrett, T., Pizzico, M., Levy, B. D., Nagel, R. L., Linsey, J. S., Talley, K. G., . . . Newstetter, W. C. (2015). A Review of University Maker Spaces.

Barro, R. J., & Lee, J.-W. (2000). Center for International Development.

Basiago, A. (1996). The search for the sustainable city in 20th century urban planning. *Environmentalist, 16*(2), 135-155.

Bates, A. K. (2003). *Ecovillages. Encyclopedia of Community: From the Village to the Virtual World. SAGE Publications, Inc.* Thousand Oaks, CA: SAGE Publications, Inc.

Bates, R., Greif, A., Levi, M., Rosenthal, J.-L., & Weingast, B. (2000). Analytic narratives revisited. *Social Science History, 24*(04), 685-696.

Bathelt, H., Malmberg, A., & Maskell, P. (2004). Clusters and knowledge: local buzz, global pipelines and the process of knowledge creation. *Progress in human geography, 28*(1), 31-56.

Battaglia, A., & Tremblay, D.-G. (2012). 22@ and the Innovation District in Barcelona and Montreal: a process of clustering development between urban regeneration and economic competitiveness. *Urban Studies Research, 2011*.

Becattini, G. (1962). *Il concetto d'industria e la teoria del valore*: P. Boringhieri.

Becattini, G. (1975). *Invito a una rilettura di Marshall*: ISEDI.

Belussi, F., & Caldari, K. (2009). At the origin of the industrial district: Alfred Marshall and the Cambridge school. *Cambridge Journal of Economics, 33*(2), 335-355.

Bennett, A. (2008). Process tracing: A Bayesian perspective. *The Oxford handbook of political methodology*, 702-721.

Bennett, A. (2010). Process tracing and causal inference. In H. E. Brady, and David Collier (Ed.), *Rethinking social inquiry: Diverse tools, shared standards*: Rowman & Littlefield Publishers.

257

Bennett, A., & Checkel, J. T. (2012). Process tracing: from philosophical roots to best practices. *Simons Papers in Security and Development, 21*, 30.

Bennett, A., & Checkel, J. T. (2014). *Process tracing*: Cambridge University Press.

Bestor, A. E. (1950). *Backwoods utopias*: University of Pennsylvania Press.

Bestor, A. E., Jr. (1948). The Evolution of the Socialist Vocabulary. *Journal of the History of Ideas, 9*(3), 259-302. Retrieved from http://www.jstor.org/stable/2707371

Bestor, A. E., Jr. (1953). Patent-Office Models of the Good Society: Some Relationships between Social Reform and Westward Expansion. *The American Historical Review, 58*(3), 505-526. Retrieved from http://www.jstor.org/stable/1843945

Birch, E. L. (1980). Radburn and the American Planning Movement The Persistence of an Idea.

Bochinski, L. B. (2016). *Alumni of Experimental Communities: Agents of Change at a Critical Time.* Prescott College.

Boix, R., & Galletto, V. (2009). Innovation and industrial districts: a first approach to the measurement and determinants of the I-district effect. *Regional Studies, 43*(9), 1117-1133.

Boix, R., Sforzi, F., & Hernández, F. (2015). Introduction: Rethinking industrial districts in the XXI Century. *Investigaciones regionales - Journal of REGIONAL RESEARCH*(32), 5-8.

Boix, R., & Trullén, J. (2010). Industrial districts, innovation and I-district effect: territory or industrial specialization? *European Planning Studies, 18*(10), 1707-1729.

Boyer, R. H. (2015). Grassroots innovation for urban sustainability: comparing the diffusion pathways of three ecovillage projects. *Environment and Planning A, 45*, 320-337.

Brabant, C. (2012, July 23, 2012). [Opportunitities for McGill].

Bracken, D. (2015, September 25, 2015). Project that will add retail, residential to RTP set to begin Jan. 1. *The News & Observer*. Retrieved from http://www.newsobserver.com/news/business/article37113903.html

Bramwell, A., Hepburn, N., & Wolfe, D. A. (2012). Growing innovation ecosystems: university-industry knowledge transfer and regional economic development in Canada. *Knowledge Synthesis Paper on Leveraging Investments in HERD. Final Report to the Social Sciences and Humanities Research Council of Canada.*

Braungart, M. (2000). Cradle to Cradle. In B. McDonough (Ed.): Mc GrawHill.

Brindley, T. (2003). Village and community: social models for sustainable urban development. *People, Places and Sustainability*, 66-82.

Brown, K. (2015). John T. Lyle Center for Regenerative Studies. . Retrieved from https://www.cpp.edu/~crs/

Brown, S. L. (2002). *Intentional community: an anthropological perspective*: SUNY Press.

Brundtland, G., Khalid, M., Agnelli, S., Al-Athel, S., Chidzero, B., Fadika, L., . . . de Botero, M. M. (1987). Our Common Future

Brusco, S. (1990). The idea of the industrial district: its genesis. *Industrial districts and inter-firm co-operation in Italy*, 10-19.

Buttel, F., Bell, M., Bunker, S., Mirata, A., Overdevest, C., Brewster, B., . . . Pellow, D. (2004). Political economy and environmental crisis. *Organization & Environment, 17*(3), 293-295.

Cal-Poly. (2015). The history of the Lyle Center - Cal Poly Pomona. Retrieved from https://www.cpp.edu/~crs/history.html

Calhoun, C. (1998). Community without propinquity revisited: Communications technology and the transformation of the urban public sphere. *Sociological Inquiry, 68*(3), 373-397.

Camisón, C., & Villar ‐ López, A. (2012). On how firms located in an industrial district profit from knowledge spillovers: adoption of an organic structure and innovation capabilities. *British journal of management, 23*(3), 361-382.

Campbell, R. (2005, March 20, 2005). Universities are the new city planners. *Boston Globe.*

Carlino, G. A., Chatterjee, S., & Hunt, R. M. (2007). Urban density and the rate of invention. *Journal of Urban Economics, 61*(3), 389-419.

Carlyle, T., & Emerson, R. W. (1884). *The Correspondence of Thomas Carlyle and Ralph Waldo Emerson, 1834-1872* (Vol. 2): Houghton, Mifflin.

Carmony, D. F., & Elliott, J. M. (1980). *New Harmony, Indiana: Robert Owen's Seedbed for Utopia*: JSTOR.

Carr, E. H., & Davies, R. W. (1961). What is history?

Cary, J. (1998). *Institutional innovation in natural resource management in Australia: The triumph of creativity over adversity.* Paper presented at the Conference "Knowledge Generation and transfer: Implications for Agriculure in the 21st Century ". University of California-Berkeley.

Castrejon Cardenas, C. (2007). *A Simple Life? the symbolic significance of environmentalism in the construction of a community: case study in the ecovillage of Las Nubes in Veracruz, Mexico.*

Ceulemans, K., & De Prins, M. (2010). Teacher's manual and method for SD integration in curricula. *Journal of Cleaner Production, 18*(7), 645-651.

Chatterji, A., Glaeser, E. L., & Kerr, W. R. (2013, April 2013). *Clusters of entrepreneurship and innovation.* Working Paper, No. 13–090. NBER Working Paper

Series, No. 19013. National Bureau of Economic Research.

Chesbrough, H. (2006). Open innovation: a new paradigm for understanding industrial innovation. *Open innovation: Researching a new paradigm*, 1-12.

Chesbrough, H. (2013). *Open business models: How to thrive in the new innovation landscape*: Harvard Business Press.

Chesbrough, H., & Bogers, M. (2014). Explicating open innovation: clarifying an emerging paradigm for understanding innovation. *New Frontiers in Open Innovation. Oxford: Oxford University Press, Forthcoming*, 3-28.

Chesbrough, H. W. (2006). *Open innovation: The new imperative for creating and profiting from technology*: Harvard Business Press.

259

Chitewere, T. (2006). *Constructing a green lifestyle: Consumption and environmentalism in an ecovillage.* (3211269 Ph.D.), State University of New York at Binghamton, Ann Arbor. ProQuest Dissertations & Theses Full Text database.

Chitewere, T. (2010). Equity in sustainable communities: exploring tools for environmental justice and political ecology. *Nat. Resources J., 50*, 315.

Chitewere, T., & Taylor, D. E. (2010). Sustainable living and community building in Ecovillage at Ithaca: The challenges of incorporating social justice concerns into the practices of an ecological cohousing community. *Research in Social Problems and Public Policy, 18*, 141-176.

Chodorkoff, D. (1995). Redefining development. *Democracy & Nature, 3*(1), 117-128. Retrieved from http://www.democracynature.org/vol3/chodorkoff_development.htm

Christensen, C. M., & Eyring, H. J. (2011). The innovative university: San Francisco: Jossey Bass.

Christian, D. L. (2003). *Creating a life together: Practical tools to grow ecovillages and intentional communities*: New Society Publishers.

Chytry, J. (2012). Walt Disney and the creation of emotional environments: interpreting Walt Disney's oeuvre from the Disney studios to Disneyland, CalArts, and the Experimental Prototype Community of Tomorrow (EPCOT). *Rethinking History, 16*(2), 259-278.

Ciao Innovation District! Menino shares Boston's innovation agenda in Italy. (2010). Retrieved from http://www.innovationdistrict.org/2010/12/02/ciao-innovation-district-menino-shares-bostons-innovation-agenda-in-italy/

Claeys, G. (2010). *The Cambridge companion to utopian literature*: Cambridge University Press.

Claeys, G. (2011). *Searching for Utopia: The history of an idea*: Thames & Hudson Limited.

Clandinin, D., Huber, J., McGaw, B., Baker, E., & Peterson, P. (2010). International encyclopedia of education: Elsevier New York, NY.

Clandinin, D. J., & Connelly, F. M. (2000). Narrative inquiry. *International encyclopedia of education*, 1-23.

Clark, J., Huang, H.-I., & Walsh, J. P. (2010). A typology of 'innovation districts': what it means for regional resilience. *Cambridge Journal of Regions, Economy and Society, 3*(1), 121-137.

Clayton, R. (1996). The logic of historical explanation: University Park, Pa.: Penn State University Press.

Cohen, B. (2006). Sustainable valley entrepreneurial ecosystems. *Business Strategy and the Environment, 15*(1), 1-14.

Cohen, W. M., & Levinthal, D. A. (1990). Absorptive capacity: A new perspective on learning and innovation. *Administrative science quarterly*, 128-152.

Collier, D. (2011). Understanding process tracing. *PS Political science and politics, 44*(4), 823.

Cortese, A. D. (2003). The critical role of higher education in creating a sustainable future. *Planning for Higher Education, 31*(3), 15-22.

Cosentini, S. (2011). Through Major EPA Grant, Tompkins County Recognized as a Climate Showcase Community. Retrieved from http://newearthliving.net/press/press-release-april-2011/

Crow, M. M. (2008). *Building an entrepreneurial university.* Paper presented at the the future of the Research University. Meeting.

Crow, M. M. (2009). *The research university as comprehensive knowledge enterprise: The reconceptualization of Arizona State University as a prototype for a new American university.* Paper presented at the Seventh Glion Colloquium, Montreux, Switzerland.

Cugurullo, F. (2013). How to build a sandcastle: An analysis of the genesis and development of Masdar City. *Journal of Urban Technology, 20*(1), 23-37.

Cullingford, C., & Blewitt, J. (2013). *The Sustainability Curriculum: The Challenge for Higher Education*: Routledge.

Cummings, M. S. (2003). *Intentional Communities and Mainstream Politics. Encyclopedia of Community: From the Village to the Virtual World. SAGE Publications, Inc.* Thousand Oaks, CA: SAGE Publications, Inc.

D'Este, P., Guy, F., & Iammarino, S. (2012). Shaping the formation of university–industry research collaborations: what type of proximity does really matter? *Journal of economic geography*.

Dawson, J. (2006). *Ecovillages: New Frontiers for Sustainability, Schumacher Briefing*: Chelsea Green Publishing.

Dawson, J. (2013). From Islands to Networks: The History and Future of the Ecovillage Movement. *Environmental anthropology engaging ecotopia: bioregionalism, permaculture, and ecovillages. New York: Berghahn Books*, 217-234.

de Oliveira Arend, C. (2013). *Reinventing the Wheel to Guide Ecovillages towards Sustainability.* Blekinge Institute of Technology.

Dedrick, J., & West, J. (2003). *Why firms adopt open source platforms: A grounded theory of innovation and standards adoption.* Paper presented at the Proceedings of the workshop on standard making: A critical research frontier for information systems.

Delong, D., & McDermott, M. (2013). Current perceptions, prominence and prevalence of sustainability in the marketing curriculum. *Marketing Management Journal, 23*(2), 101-116.

Denehie, E. S. (1923). The Harmonist Movement in Indiana. *Indiana Magazine of History, 19*(2), 188-200. Retrieved from http://www.jstor.org/stable/27786079

Diamond, J. (2005). *Collapse: How societies choose to fail or succeed*: Penguin.

DiBona, C., & Ockman, S. (1999). *Open sources: Voices from the open source revolution*: O'Reilly Media, Inc.

Disney, W. (Writer). (1966). Project florida: a whole new disneyworld.

Doan, A., Ramakrishnan, R., & Halevy, A. Y. (2011). Crowdsourcing systems on the world-wide web. *Communications of the ACM, 54*(4), 86-96.

DOE. (2015). DOE Releases Common Definition for Zero Energy Buildings, Campuses, and Communities. Retrieved from http://energy.gov/eere/buildings/articles/doe-releases-common-definition-zero-energy-buildings-campuses-and

Dressler, L. (2006). *Consensus Through Conversations: How to Achieve High-Commitment Decisions*: Berrett-Koehler Publishers.

Durrett, C. (2015). Retrieved from http://www.cohousingco.com/bios/

Eagan, D. J., & Orr, D. W. (1992). *Campus and environmental responsibility*: Jossey-Bass.

Eccles, R. G., & Krzus, M. P. (2010). *One report: Integrated reporting for a sustainable strategy*: John Wiley & Sons.

EcoInnovation district initiative. (2014). Retrieved from http://www.csndc.com/about.php

EcoInnovation District Uptown Oakland. (2016). Retrieved from http://www.ecoinnovationdistrict.org/

Ed, P. (2016). Innovation Square. Retrieved from http://innovationsquare.ufl.edu/inspiration-hall

Ehrenberg, R. G. (2012). American higher education in transition. *The Journal of Economic Perspectives, 26*(1), 193-216.

Eilperin, J. (2005). Colleges Compete to Shrink Their Mark on the Environment. *Washington Post, 26*.

Einstein, A. (1946, May 25, 1946). Atomic Education Urged by Einstein. *New York Times*.

Eisenhardt, K. M. (1989). Building theories from case study research. *Academy of management review, 14*(4), 532-550.

Elkington, J. (1997). Cannibals with forks. *The triple bottom line of 21st century*.

Enright, M. J. (2003). Regional clusters: what we know and what we should know *Innovation clusters and interregional competition* (pp. 99-129): Springer.

Ergas, C. (2010). A Model of Sustainable Living: Collective Identity in an Urban Ecovillage. *Organization & Environment, 23*(1), 32.

Ernst, H. C. (1904). *Some fermentations in medical education*.

Etzkowitz, H. (2013). Silicon Valley at risk? Sustainability of a global innovation icon: An introduction to the Special Issue. *Social Science Information, 52*(4), 515-538.

Etzkowitz, H. (2014). Making a humanities town: knowledge-infused clusters, civic entrepreneurship and civil society in local innovation systems. *Triple Helix, 2*(1), 1-22.

Farr, D. (2011). *Sustainable urbanism: urban design with nature*: John Wiley & Sons.

Feld, B. (2012). *Startup communities: Building an entrepreneurial ecosystem in your city*: John Wiley & Sons.

Ferrer-Balas, D., Lozano, R., Huisingh, D., Buckland, H., Ysern, P., & Zilahy, G. (2010). Going beyond the rhetoric: system-wide changes in universities for sustainable societies. *Journal of Cleaner Production, 18*(7), 607-610.

FIC. (2016). Fellowship of Intentional Community. Retrieved from http://www.ic.org/

Fields, K. (2012). The 2012 Evergreen Awards: TerraHaus at Unity College. Retrieved from http://www.architectmagazine.com/awards/the-2012-evergreen-awards-terrahaus-at-unity-college_o

Fischetti, D. M. (2008). *Building resistance from home: Ecovillage at Ithaca as a model of sustainable living.* University of Oregon.

Fischler, R. (2013). *QI meeting.* Retrieved from School of Urban Planning:

Fleming, S. (2016). Innovation U 2.0. Retrieved from http://academicvc.com/2014/04/01/innovation-u-2-0/

Fling, F. M. (1899). *Outline of historical method*: JH Miller.

Flint, A. (2016). Are 'innovation districts' right for every city? *City Lab.* Retrieved from http://www.citylab.com/tech/2016/04/are-innovation-districts-right-for-every-city/480534/ - disqus_thread

Florida, R. (2014). The Creative Class and Economic Development. *Economic development quarterly, 28*(3), 196-205.

Florida, R. L. (2002). *The rise of the creative class: and how it's transforming work, leisure, community and everyday life*: Basic books.

Flynn, W. J., & Vredevoogd, J. (2010). The Future of Learning: 12 Views on Emerging Trends in Higher Education. *Planning for Higher Education, 38*(2), 5-10.

Fogarty, R. S. (1980). *Dictionary of American communal and utopian history*: Westport, Conn.: Greenwood Press.

Frank, P. (1951). *Einstein, His Life and Times.*

Frederick, H. H. (2015). The role of universities as entrepreneurship ecosystems in the era of climate change: A new theory of entrepreneurial ecology. *Jurnal Intelek, 6*(2).

Fullan, M. (2006). The future of educational change: System thinkers in action. *Journal of educational change, 7*(3), 113-122.

Fuller, R. B. (1982). *Critical path*: Macmillan.

Gadotti, M. <Education for Sustainable Development-1.pdf>.

Gadotti, M. (2010). Reorienting education practices towards sustainability. *Journal of Education for Sustainable Development, 4*(2), 203-211.

Galbraith, S. (2012). Watch Out Silicon Valley! Colorado Primed to Emerge as the Next Hub of Innovation and Entrepreneurship. *Forbes.* Retrieved from http://www.forbes.com/sites/sashagalbraith/2012/12/17/watch-out-silicon-valley-colorado-primed-to-emerge-as-the-next-hub-of-innovation-and-entrepreneurship/

Galea, C. (2007). *Teaching business sustainability*: Greenleaf.

Garforth, L. (2009). No Intentions? Utopian Theory After the Future. *Journal for Cultural Research, 13*(1), 5-27.

Gassmann, O., Enkel, E., & Chesbrough, H. (2010). The future of open innovation. *R&d Management, 40*(3), 213-221.

George, A., & Bennett, A. (1979). *Case studies and theory development*: Free Press.

Gibson, D. V., Foss, L., & Hodgson, R. (2014). Institutional Perspectives in Innovation Ecosystem Development *Moderne Konzepte des organisationalen Marketing* (pp. 61-75): Springer.

Giuffrida, G., Clark, J. J., & Cross, S. E. (2015). *Putting Innovation in Place: Georgia Tech's Innovation Neighbourhood of 'Tech Square'.* Paper presented at the Presented at the European Conference on Innovation and Entrepreneurism.

Gladman, K. J. (2014). *Partnerships for Sustainability: Eco-Collaboration between Higher Education and Ecovillages.* University of Minnesota.

Glenn, T. a. R., Brooks. (2016, April 18, 2016). The Chattanooga Story.

Goldman, R., & Gabriel, R. P. (2005). *Innovation happens elsewhere: Open source as business strategy*: Morgan Kaufmann.

Goldsmith, E., Allen, R., Allaby, M., Davoll, J., & Lawrence, S. (1972). *A blueprint for survival*: Houghton Mifflin Boston.

Goldsmith, Z. (2007). You have been warned-Actor and environmental activist Leonardo DiCaprio's new film The 11th Hour is a harrowing look into the future. Zac Goldsmith interviews the man of the hour. *Ecologist, 37*(7), 28-33.

Gomez, C. (2014). Changing continuing education landscape holds new challenges.

Goodman, E. J., Bamford, J., & Saynor, P. (1989). *Small firms and industrial districts in Italy*: Taylor & Francis.

Goodwin, B. (2012). *The Philosophy of Utopia*: Routledge.

Gottschalk, L. R. (1966). *Understanding history: A primer of historical method*.

Graham, R. (2013). Technology innovation ecosystem benchmarking study: key findings from phase 1. *no. January*, 24.

Granovetter, M. (1983). The strength of weak ties: A network theory revisited. *Sociological theory, 1*(1), 201-233.

Grant, L. K. (2010). Sustainability: from excess to aesthetics. *Behavior and Social Issues, 19*, 5-45.

Green Building. (2016).   Retrieved from https://archive.epa.gov/greenbuilding/web/html/about.html

Greenberg, D. (2013a). Academia's Hidden Curriculum and Ecovillages as Campuses for Sustainability Education. *Environmental Anthropology Engaging Ecotopia: Bioregionalism, Permaculture, and Ecovillages, 17*, 269.

Greenberg, D. (2013b). Daniel Greenberg.   Retrieved from http://www.globalecovillages.org/profile/DanielGreenberg?xg_source=activity

Greenberg, S. R. (2015). *Austin Anchors & The Innovation Zone: Building Collaborative Capacity*. Retrieved from LBJ School of Public Affairs:

Greiger, R. I. (2004). Knowledge and Money: research universities and the paradox of paradox or marketplace: Palo Alto: Stanford University Press.

Grierson, D. (2003). Arcology and Arcosanti: towards a sustainable built environment. *Electronic Green Journal, 1*(18).

Hall, P. A. (2006). Systematic process analysis: when and how to use it. *European Management Review, 3*(1), 24-31.

Hamdouch, A. (2007). *Innovation clusters and networks: a critical review of the recent literature.* Paper presented at the 19th EAEPE Conference.

Hamdouch, A. (2008). Conceptualizing innovation clusters and networks.

Hardin, G. (1968). The tragedy of the commons. *Science, 162*(3859), 1243-1248.

Hardy, D. (2000). *Utopian England: community experiments, 1900-1945*: Psychology Press.

Hardy, D. (2005). Garden cities: practical concept, elusive reality. *Journal of Planning History, 4*(4), 383-391.

Harlow, J., Golub, A., & Allenby, B. (2011). A Review of Utopian Themes in Sustainable Development Discourse. *Sustainable Development*.

Hart, T. A., Fox, C. J., Ede, K. F., & Korstad, J. (2015). Do, but don't tell: the search for social responsibility and sustainability in the websites of the top-100 US MBA programs. *International Journal of Sustainability in Higher Education, 16*(5), 706-728.

Hartley, J., Potts, J., & MacDonald, T. (2012). Creative city index. *Cultural Science, 5*(1).

Harvey, D. (1989). From managerialism to entrepreneurialism: the transformation in urban governance in late capitalism. *Geografiska Annaler. Series B. Human Geography*, 3-17.

Hawken, P. (1988). *Growing a business*: Simon and Schuster.

Hawken, P. (1993a). The ecology of commerce: A declaration of sustainability. *New York: Harpers Business*(59), 54-61.

Hawken, P. (1993b). *The ecology of commerce: how business can save the planet*: Weidenfeld and Nicolson.

Hawken, P. (2007). *Blessed unrest: How the largest movement in the world came into being, and why no one saw it coming*: Penguin.

Hawken, P. (2010a). *The ecology of commerce: A declaration of sustainability*: Harper Business.

Hawken, P. (2010b). *The ecology of commerce: A declaration of sustainability*: Harper Business.

Hawken, P., & Herr, M. (1975). *The magic of Findhorn*: Souvenir Press.

Hayden, D. (1976). *Seven American utopias: the architecture of communitarian socialism, 1790-1975*: MIT Press Cambridge.

Heck, J. (2015). Airbus details move to Wichita State's innovation campus. *Wichita Business Journal*. Retrieved from http://www.bizjournals.com/wichita/news/2015/03/24/airbus-details-move-to-wichita-state-s-innovation.html

Heinberg, R. (2010). *Peak everything: waking up to the century of declines*: New Society Publishers.

Hervas-Oliver, J.-L., Gonzalez, G., Caja, P., & Sempere-Ripoll, F. (2015). Clusters and industrial districts: Where is the literature going? Identifying emerging sub-fields of research. *European Planning Studies, 23*(9), 1827-1872.

Hicks, G. L. (2001). *Experimental Americans: Celo and utopian community in the twentieth century*: University of Illinois Press.

Hippel, E. v., & Krogh, G. v. (2003). Open source software and the "private-collective" innovation model: Issues for organization science. *Organization Science, 14*(2), 209-223.

Hogarth, R. M. (1974). Process tracing in clinical judgment. *Behavioral Science, 19*(5), 298-313.

Hong, S., & Vicdan, H. (2015). Re-imagining the utopian: Transformation of a sustainable lifestyle in ecovillages. *Journal of Business Research*.

Howard, E. (1902). Garden cities of tomorrow. *London, itd: Faber and Faber*.

Howard, S. E., Mumford, L., & Osborn, S. F. J. (1946). *Garden Cities of To-morrow*: Faber & Faber.

Hoxie, R. F. (1906). Historical Method vs. Historical Narrative. *The Journal of Political Economy, 14*(9), 568-572.

Huggett, B. (2014). Reinventing tech transfer. *Nature biotechnology, 32*(12), 1184-1191. Retrieved from http://www.nature.com/nbt/journal/v32/n12/pdf/nbt.3085.pdf

Hulme, P. E. (2014). Editorial: Bridging the knowing–doing gap: know‐who, know‐what, know‐why, know‐how and know‐when. *Journal of Applied Ecology, 51*(5), 1131-1136.

Hwang, V. W., & Horowitt, G. (2012). *The Rainforest: The secret to building the next Silicon Valley*: Regenwald Los Altos.

Ickes, W. (1993). Empathic accuracy. *Journal of personality, 61*(4), 587-610.

Initiative, C. (2000). The earth charter. Retrieved from http://earthcharter.org/

International Integrated Reporting Council (2016). Retrieved from http://www.iasplus.com/en/resources/sustainability/iirc

Isenberg, D. (2011). The entrepreneurship ecosystem strategy as a new paradigm for economic policy: Principles for cultivating entrepreneurship. *Presentation at the Institute of International and European Affairs*.

Jackson, H., & Mead, M. (1998). *What is an Ecovillage?* Paper presented at the Gaia Trust Education Seminar.

Jaffe, A. B., Trajtenberg, M., & Henderson, R. (1993). Geographic localization of knowledge spillovers as evidenced by patent citations. *the Quarterly journal of Economics*, 577-598.

Jarvis, H., & Bonnett, A. (2013). Progressive Nostalgia in Novel Living Arrangements: A Counterpoint to Neo-traditional New Urbanism? *Urban Studies*.

Jay, J., & Gerard, M. (2015). Accelerating the Theory and Practice of Sustainability-Oriented Innovation. *Social Science Research Network*.

Jobs, S. (2005). *Stanford commencement speech*.

Kallushi, A., Harris, J., Miller, J., Johnston, M., & Ream, A. (2012). *Think bigger: net-zero communities*. Paper presented at the Proceedings of ACEEE.

Kanter, R. M. (1968). Commitment and social organization: A study of commitment mechanisms in utopian communities. *American sociological review*, 499-517.

Kanter, R. M. (1972). *Commitment and community: Communes and utopias in sociological perspective* (Vol. 36): Harvard University Press.

Kanter, R. M. (2000). When a thousand flowers bloom: Structural, collective, and social conditions for innovation in organization. *Entrepreneurship: the social science view*, 167-210.

Kanter, R. M. (2002). Creating the culture for innovation. *Leading for innovation and organizing for results*, 73-85.

Kanter, R. M. (2003). *Challenge of organizational change: How companies experience it and leaders guide it*: Simon and Schuster.

Kanter, R. M. (2006). Innovation: the classic traps. *Harvard Business Review, 84*(11), 72-83, 154.

Kanter, R. M. (2012). Enriching the ecosystem. *Harvard Business Review, 90*(3), 140-147.

Katz, B. (2015). Retrieved from http://www.brookings.edu/research/papers/2015/06/24-one-year-innovation-districts-katz-vey-wagner

Katz, B., & Bradley, J. (2013). *The metropolitan revolution: How cities and metros are fixing our broken politics and fragile economy*: Brookings Institution Press.

Katz, B. a. W., Julie. (2014, May 2014). *The Rise of Innovation Districts: A New Geography of Innovaton in America*

Katz, B., V., Jennifer; Wagner, Julie. (2015). One year after: observations on the rise of innovation districts. Retrieved from http://www.brookings.edu/research/papers/2015/06/24-one-year-innovation-districts-katz-vey-wagner

Kibert, C. J. (2004). Green buildings: an overview of progress. *Journal of Land Use & Environmental Law*, 491-502.

Kim, M. (2013). *Spatial qualities of innovation districts: how Third Places are changing the innovation ecosystem of Kendall Square.* Massachusetts Institute of Technology.

Kiron, D., Kruschwitz, N., Reeves, M., & Goh, E. (2013). The benefits of sustainability-driven innovation. *MIT SLOAN MANAGEMENT REVIEW, 54*(2), 69.

Klein, J. T. (1990a). *Interdisciplinarity: History, theory, and practice*: Wayne State University Press.

Klein, J. T. (1990b). Interdisciplinary resources: A bibliographical reflection. *Issues in Integrative Studies, 8*, 35-67.

Klemmer, P., Lehr, U., & Loebbe, K. (1999). Environmental Innovation. Volume 3 of publications from a Joint Project on Innovation Impacts of Environmental Policy Instruments. *Synthesis Report of a project commissioned by the German Ministry of Research and Technology, Analytica-Verlag, Berlin.*

Klewitz, J., & Hansen, E. G. (2014). Sustainability-oriented innovation of SMEs: a systematic review. *Journal of Cleaner Production, 65*, 57-75.

Kliewer, J. R. (1999). *The Innovative Campus: Nurturing the Distinctive Learning Environment*: ERIC.

Knoedler, J. T. (2015). Going to College on My iPhone. *Journal of Economic Issues, 49*(2), 329-354.

Kozeny, G. (2003). Intentional Communities and Daily Life. *Encyclopedia of Community: From the Village to the Virtual World, 2*, 685-689.

Kraftl, P. (2007). Utopia, performativity, and the unhomely. *Environment and Planning D: Society and Space, 25*(1), 120-143.

Landry, C. (2005). Lineages of the creative city. *Creativity and the City, Netherlands Architecture Institute.*

Lang, D. J., Wiek, A., Bergmann, M., Stauffacher, M., Martens, P., Moll, P., . . . Thomas, C. J. (2012). Transdisciplinary research in sustainability science: practice, principles, and challenges. *Sustainability science, 7*(1), 25-43.

Laperche, B., Sommers, P., & Uzunidis, D. (2010). *Innovation Networks and Clusters: The Knowledge Backbone*: Peter Lang.

Lau, A. (2012). Masdar City: A model of urban environmental sustainability. *Social Sciences*.

Leadbetter, C. (2008). We-think: The power of mass creativity: London: Profile Books Ltd.

Lee, S. M., Olson, D. L., & Trimi, S. (2012). Co-innovation: convergenomics, collaboration, and co-creation for organizational values. *Management Decision, 50*(5), 817-831.

Leon, N. (2008). Attract and connect: The 22@ Barcelona innovation district and the internationalisation of Barcelona business. *Innovation, 10*(2-3), 235-246.

Lerner, J., & Triole, J. (2000). *The simple economics of open source.* Retrieved from

Leslie, S. W., & Kargon, R. H. (1996). Selling Silicon Valley: Frederick Terman's model for regional advantage. *Business History Review, 70*(04), 435-472.

Levitas, R. (2010). *The concept of utopia* (Vol. 3): Peter Lang.

Leydesdorff, L. (1988). *A Triple Helix of University-industry-government Relations: The Future Location of Research?* : State University of New York.

Linebaugh, P. (2008). *The Magna Carta manifesto: liberties and commons for all*: Univ of California Press.

Link, A. N. (1995). *A generosity of spirit: The early history of the Research Triangle Park*: Research Triangle Foundation of North Carolina.

Link, A. N., & Scott, J. T. (2005). Opening the ivory tower's door: An analysis of the determinants of the formation of US university spin-off companies. *Research policy, 34*(7), 1106-1112.

Link, A. N., & Scott, J. T. (2006). US university research parks. *Journal of Productivity Analysis, 25*(1-2), 43-55.

Link, A. N., & Scott, J. T. (2007). The economics of university research parks. *Oxford Review of Economic Policy, 23*(4), 661-674.

Link, A. N., & Scott, J. T. (2011). Research, science, and technology parks: Vehicles for technology transfer. *Department of Economics-Working Paper Series. The University of North Carolina: Greensboro*, 11-22.

Litan, R. E., Mitchell, L., & Reedy, E. (2007). The university as innovator: Bumps in the road. *Issues in Science and Technology, 23*(4), 57-66.

Litfin, K. (2007). The global ecovillage movement as a holistic knowledge community. *Reinventing the Future*.

Litfin, K. T. (2014). *Ecovillages: Lessons for sustainable community*: John Wiley & Sons.

Lockyer, J. (2009). From Developmental Communalism to Transformative Utopianism: An Imagined Conversation with Donald Pitzer. *Communal Societies, 29*(1), 1-14.

Lockyer, J., & Veteto, J. R. (2013). Environmental Anthropology Engaging Ecotopia. *Environmental Anthropology Engaging Ecotopia: Bioregionalism, Permaculture, and Ecovillages, 17*, 1.

Lockyer, J. P. (2007). *Sustainability and Utopianism: An Ethnography of Cultural Critique in Contemporary Intentional Communities.* University of Georgia.

Loezer, L. (2011). *Enhancing Sustainability at the Community Level: Lessons from American EcoVillages.* University of Cincinnati.

Longhurst, N. (2015). Towards an 'alternative' geography of innovation: Alternative milieu, socio-cognitive protection and sustainability experimentation. *Environmental Innovation and Societal Transitions, 17,* 183-198.

Lowell, A. L. (1904, June 1904). Dormitories and College Life. *Harvard Graduates' Magazine, VII,* 527.

Lower, D. E. (2012). *St. Louis innovation district tax increment financing redevelopment plan.* . Retrieved from https://www.stlouis-mo.gov/government/departments/sldc/documents/cortex-tif-st-louis-innovation-district.cfm

MacGregor, J., & Smith, B. L. (2005). Where Are Learning Communities Now?: National Leaders Take Stock. *About Campus, 10*(2), 2-8.

Mahoney, J. (2015). Process Tracing and Historical Explanation. *Security Studies, 24*(2), 200-218.

Majid, S. (2014). *Austin Innovation Zone: The New Economic Geography.* Retrieved from

Mannheim, K. (1936). Ideology and Utopia: An Introduction to the Sociology of Knowledge, trans. Louis Wirth and Edward Shils. *New York: A Harvest Book, Harcourt, Brace and World.*

Mannheim, S. (2012). *Walt Disney and the quest for community*: Ashgate Publishing, Ltd.

Marin County voters reject 'solar village'. (1979, November 7, 1979). *Eugene Register-Guard.*

Mark, C. (1991). *This is not my beautiful house: cohousing as an alternative American Dream.* Massachusetts Institute of Technology.

Marks, L. M., Katie. (2012). *Cost- Benefit Analysis of the Impact of the St. Louis Innovation District TIF Redevelopment Plan.* Retrieved from

Markusen, A. (1996). Sticky places in slippery space: a typology of industrial districts. *Economic geography,* 293-313.

Marshall, A. (1920). *Principles of economics: an introductory volume.*

Martin, J. (2012). *The sustainable university: Green goals and new challenges for higher education leaders*: JHU Press.

Martin, R., & Sunley, P. (2003). Deconstructing clusters: chaotic concept or policy panacea? *Journal of economic geography, 3*(1), 5-35.

Mason, C., & Brown, R. (2014). Entrepreneurial ecosystems and growth oriented entrepreneurship. *Final Report to OECD, Paris.*

Maxwell, J. A. (1992). Understanding and validity in qualitative research. *Harvard educational review, 62*(3), 279-301.

Maxwell, J. A. (2012). *A realist approach for qualitative research*: Sage.

Mayumi, K. (2002). *The origins of ecological economics: the bioeconomics of Georgescu-Roegen*: Routledge.

Mazur, L. F., Denise (2015). Is "resilience" the new sustainababble? Retrieved from http://grist.org/article/is-resilience-the-new-sustainababble/

McCamant, K., & Durrett, C. (1994). Cohousing. *A Contemporary Approach to Housing Ourselves, 2*.

McLaughlin, C., & Davidson, G. (1985). *Builders of the dawn: Community lifestyles in a changing world*: Stillpoint Pub.

Meadows, D. (1999). Leverage points: Places to intervene in a system. *The Sustainability Institute*.

Meiklejohn, A. (1932). *The experimental college*: Univ of Wisconsin Press.

Meltzer, G. (2005). *Sustainable community: Learning from the cohousing model*: Trafford on Demand Pub.

Metcalf, B., & Christian, D. L. (2003). Overview of Intentional Communities.

Miller, J. (2015). *Uptown EcoInnovation District* Retrieved from

Miller, T. (1992). The roots of the 1960s communal revival. *American Studies*, 73-93.

Miller, T. (2010). A Matter of Definition: Just What Is an Intentional Community? *Communal Societies, 30*(1), 1-15.

Miller, V., Washburn, A., Norby, M., Banset, E., & Klucas, G. (2015). Research at the University of Nebraska-Lincoln: 2014-2015 Report.

Mini Town Hall Innovation Discussions. (2015). Retrieved from http://webs.wichita.edu/?u=president&p=/mth_listing/

MIT. (2016). Mit open courseware site stats. Retrieved from http://ocw.mit.edu/about/site-statistics/

Moore, W. B. (2012). Sustainability reporting.

Morisson, A. (2014). Innovation districts: an investigation of the replication of the 22@ Barcelona's Model in Boston.

Morisson, A. (2015). *Innovation districts: a toolkit for urban leaders*: CreateSpace Independent Publishing Platform.

Morrison, E. (2013). Distinguishing innovation ecosystems from entrepreneurial ecosystems. Retrieved from http://www.edmorrison.com/distinguishing-innovation-ecosystems-from-entrepreneurial-ecosystems/

Morrison, E. (2014). *Universities as Anchors for Regional Innovation Ecosystems*. Retrieved from

Moss Kanter, R. (2012). Enriching the Ecosystem-A four-point plan for linking innovation, enterprises, and jobs. *Harvard Business Review*, 140.

Muir, J., Limbaugh, R. H., & Lewis, K. E. (1985). *The John Muir Papers, 1858-1957*: Chadwyck-Healey.

Mulder, K., Costanza, R., & Erickson, J. (2006). The contribution of built, human, social and natural capital to quality of life in intentional and unintentional communities. *Ecological Economics, 59*(1), 13-23.

Mundigo, A. I., & Crouch, D. P. (1977). The city planning ordinances of the laws of the indies revisited: Part I: Their philosophy and implications. *Town Planning Review, 48*(3), 247.

Muro, M., & Katz, B. (2010). *The new 'cluster moment': how regional innovation clusters can foster the next economy*: September.

Muscio, A. (2006). Patterns of innovation in industrial districts: an empirical analysis. *Industry and Innovation, 13*(3), 291-312.

Muscio, A., Quaglione, D., & Scarpinato, M. . (2012). The effects of universities' proximity to industrial districts on university–industry collaboration. *China Economic Review, 23*(3), 639-650.

Mychajluk, L. H. (2014). *Building capacity to live and work together at an ecovillage in support of sustainable community: a case study Masters of Arts 2014.* University of Toronto.

Nachman, B. R. (2014). Epcot's Evolution: Disney's Ultimate World's Fair of Technological and Cultural Synergy.

Nations, U. (2015). *World population prospects: the 2015 revision.* Retrieved from http://www.un.org/en/development/desa/publications/world-population-prospects-2015-revision.html

Network, S. V. (2004). *Joint Venture's Index of Silicon Valley*: Joint Venture, Silicon Valley Network.

Newell, W. H. (2001). A theory of interdisciplinary studies. *Issues in Integrative Studies, 19*(1), 1-25.

Nicolescu, B. (2005). *Transdisciplinarity: past, present and future.* Paper presented at the Congresso Mundial De.

Norgaard, R. B. (1984). Coevolutionary development potential. *Land Economics, 60*(2), 160-173.

Norkus, Z. (2005). Mechanisms as miracle makers? The rise and inconsistencies of the "mechanismic approach" in social science and history. *History and Theory, 44*(3), 348-372.

O'mara, M. (2011). Silicon Valleys: here, there, and everywhere. *Boom: A Journal of California, 1*(2), 75-81.

O'Mara, M. P. (2015). *Cities of Knowledge: Cold War Science and the Search for the Next Silicon Valley: Cold War Science and the Search for the Next Silicon Valley*: Princeton University Press.

Oh, D.-S., Phillips, F., Park, S., & Lee, E. (2016). Innovation ecosystems: A critical examination. *Technovation.*

Orlando, M. J., & Verba, M. (2005). Do only big cities innovate? Technological maturity and the location of innovation. *Economic Review-Federal Reserve Bank of Kansas City, 90*(2), 31.

Orr, D. W. (2002). *The nature of design: ecology, culture, and human intention*: Oxford University Press.

Orr, D. W. (2004). *Earth in mind: On education, environment, and the human prospect*: Island Press.

Orr, D. W. (2012). *The last refuge: Patriotism, politics, and the environment in an age of terror*: Island Press.

Ostrom, E., Burger, J., Field, C. B., Norgaard, R. B., & Policansky, D. (1999). Revisiting the commons: local lessons, global challenges. *Science, 284*(5412), 278-282.

Ostrom, E., Walker, J., & Gardner, R. (1992). Covenants with and without a Sword: Self-governance Is Possible. *American political science Review, 86*(02), 404-417.

Pappano, L. (2012). The Year of the Mooc. Retrieved from http://www.nytimes.com/2012/11/04/education/edlife/massive-open-online-courses-are-multiplying-at-a-rapid-pace.html?pagewanted=all&_r=1&

Patagonia. (2015). Patagonia Company Info: our mission statement. Retrieved from http://www.patagonia.com/us/patagonia.go?assetid=2047

Pazzanese, C. (2014). Renewing Urban Renewal Retrieved from http://news.harvard.edu/gazette/story/2014/07/renewing-urban-renewalPentland, A. S. (2013). The data-driven society. *Scientific American, 309*(4), 78-83.

Perens, B. (1999). The open source definition. *Open sources: voices from the open source revolution*, 171-188.

Pfeifer, J., & Sutton, R. (2000). The knowing-doing gap. *Harvard, USA*.

Pickerill, J. (2015). Building the commons in eco-communities. *Space, Power and the Commons: The Struggle for Alternative Futures*, 31.

Pink, D. H. (2011). *Drive: The surprising truth about what motivates us*: Penguin.

Piore, M. J., & Sabel, C. F. (1984). *The second industrial divide: possibilities for prosperity*: Basic books.

Pitzer, D. E. (1989). Developmental communalism: An alternative approach to communal studies. *Utopian thought and communal experience*(24), 68.

Polak, F. L. (1961). The Image of the Future, 2 vols. *New York: Oceana, 50*.

Ponterotto, J. G. (2006). Brief note on the origins, Evolution, and meaning of the qualitative research concept Thick Description. *The Qualitative Report, 11*(3), 538-549.

Ponzini, D., & Rossi, U. (2010). Becoming a creative city: The entrepreneurial mayor, network politics and the promise of an urban renaissance. *Urban Studies, 47*(5), 1037-1057.

Porter, E. (2014). A Smart Way to Skip College in Pursuit of a Job Udacity-AT&T 'NanoDegree'Offers an Entry-Level Approach to College. *The New York Times Economy*.

Porter, M. E. (1998). *Clusters and the new economics of competition* (Vol. 76): Harvard Business Review.

Porter, M. E. (2008a). *Competitive advantage: Creating and sustaining superior performance*: Simon and Schuster.

Porter, M. E. (2008b). *On competition*: Harvard Business Press.

Porter, M. E. (2010). *Anchor institutions and urban economic development: From community benefit to shared value.* Paper presented at the Inner City Economic Forum Summit 2010.

Porter, M. E. (2014). U.S. Cluster Mapping Project. *Institute for Strategy and Competitiveness.* Retrieved from http://clustermapping.us/

Prahalad, C. K., & Ramaswamy, V. (2004). Co-creation experiences: The next practice in value creation. *Journal of interactive marketing, 18*(3), 5-14.

Price, A., & Delbridge, R. (2015). *Social science parks: society's new super-labs.* Retrieved from

A programme of urban, economic and social transformation. (2012).

A programme of urban, economic and social transformation. (2012). Retrieved from

Psillos, S. (2000). The present state of the scientific realism debate. *British Journal for the Philosophy of Science, 51*(4), 706.

Puddu, S., & Zuddas, F. (2013). Towards a city of innovation: the spatial regotiation between campus and city quarter.

Pyke, F., Becattini, G., & Sengenberger, W. (1990). *Industrial districts and inter-firm co-operation in Italy*: International Institute for Labour Studies Geneva.

Qingji, L. C. S. (2002). Ecological Thinking and Its Inspiration of the New Urbanism *Urban Research, 1*, 013.

Quartier de l'innovation: a joint vision for a prosperous future. (2014, 2014).

Quintas, P., Wield, D., & Massey, D. (1992). Academic-industry links and innovation: questioning the science park model. *Technovation, 12*(3), 161-175.

Raasch, C., Lee, V., Spaeth, S., & Herstatt, C. (2013). The rise and fall of interdisciplinary research: The case of open source innovation. *Research policy, 42*(5), 1138-1151.

Rainwater, B. (2013). Cities as Labs.   Retrieved from http://www.aia.org/practicing/AIAB100112

Rainwater, B. (2014). Building the new relationship infrastructure - innovation districts take off.

Raymond, E. S. (1999). The cathedral and the bazaar: Musings on Linux and open source from an accidental revolutionary. Sebastapol. *CA: O'Reilly BC Associates.*

Redden, E. (2014). Cutting ties. *Inside higher ed.* Retrieved from https://www.insidehighered.com/news/2014/01/28/study-abroad-provider-living-routes-close-after-losing-its-affiliation-umass-amherst

Rennings, K. (2000). Redefining innovation—eco-innovation research and the contribution from ecological economics. *Ecological Economics, 32*(2), 319-332.

Rice, M. P., Fetters, M. L., & Greene, P. G. (2014). University-based entrepreneurship ecosystems: a global study of six educational institutions. *International Journal of Entrepreneurship and Innovation Management, 18*(5-6), 481-501.

Rilke, R. M., & Burnham, J. (1993). Letters to a Young Poet. 1934. *Trans. MD Herter Norton. New York: WW Norton & Co.*

Rockström, J. (2009). Planetary Boundaries. *Nature, 461*, 472-475.

Rogerson, C. (1993). *Industrial districts: Italian experience, South African policy issues.* Paper presented at the Urban Forum.

Ross, C. (2014, January 10, 2014). Office rents soaring in innovation district. *The Boston Globe*. Retrieved from http://www.bostonglobe.com/business/2014/01/10/rents-soaring-city-innovation-district/nqeKNcRiLJiyjKEEGog8GP/story.html

Ruef, M. (2002). Strong ties, weak ties and islands: structural and cultural predictors of organizational innovation. *Industrial and Corporate Change, 11*(3), 427-449.

Russell, J. (2014, June 20, 2014). Density boondoogles: innovation districts. *Pacific Standard Magazine.* Retrieved from http://www.psmag.com/books-and-culture/density-boondoggles-innovation-districts-84034

273

Sanchez-Gordon, S., & Lujan-Mora, S. (2015). *An ecosystem for corporate training with accessible MOOCs and OERs.* Paper presented at the MOOCs, Innovation and Technology in Education (MITE), 2015 IEEE 3rd International Conference.

Sandelin, J. (2004). *The Story of the Stanford Industrial/Research Park.* Paper presented at the Paper delivered to the International Forum of University Science Park Beijing, China.

Sargent, L. T. (1994). The three faces of utopianism revisited. *Utopian Studies*, 1-37.

Sargent, L. T. (2012). Theorizing Intentional Community in the Twenty-First Century." *The Communal Idea in the 21 st Century* (pp. 53-72).

Sargisson, L. (2002). *Utopian bodies and the politics of transgression*: Routledge.

Sargisson, L. (2004). *Utopia and Intentional Communities.* Paper presented at the Draft paper for discussion at the ECPR Conference, Uppsala.

Scerri, A., & Holden, M. (2014). Ecological modernization or sustainable development? Vancouver's greenest city action plan: The city as 'manager'of ecological restructuring. *Journal of Environmental Policy & Planning, 16*(2), 261-279.

Schafer, W. J. (1978). Looking Backward: Early American Radicalism and Utopian Visions. *Minnesota Review, 11*(1), 60-72.

Schehr, R. C. (1997). *Dynamic Utopia: Establishing intentional communities as a new social movement*: Greenwood Publishing Group.

Selingo, J. (2013). College unbound. *New Harvest, 256*.

Selingo, J. J. (2013). *College (un) bound: The future of higher education and what it means for students*: Houghton Mifflin Harcourt.

Selingo, J. J. (2014). Location, location, location. Urban hot spots are the place to be. *The Chronical of Higher Education*.

Sforzi, F. (2015). Rethinking the industrial district: 35 years later. *Investigaciones Regionales*(32), 11.

Sharifi, A. (2016). From Garden City to Eco-urbanism: The quest for sustainable neighborhood development. *Sustainable Cities and Society, 20*, 1-16.

Shearmur, R. (2012). Are cities the font of innovation? A critical review of the literature on cities and innovation. *Cities, 29*, S9-S18.

Shearmur, R., & Doloreux, D. (2016). How open innovation processes vary between urban and remote environments: slow innovators, market-sourced information and frequency of interaction. *Entrepreneurship & Regional Development*, 1-21.

Shushok Jr, F., Penven, J., Stephens, R., & Keith, C. (2013). The Past, Present, and Future of Residential Colleges: Looking Back at S. Stewart Gordon's" Living and Learning in College". *Journal of College and University Student Housing, 39*(2), 114-127.

Siles, D. (2015). Quartier de l'innovation (QI) de Montréal: an innovation ecosystem in the heart of Montreal. Retrieved from http://quartierinnovationmontreal.com/en/discover-qi/

Simmie, J., & Hart, D. (1999). Innovation projects and local production networks: a case study of Hertfordshire. *European Planning Studies, 7*(4), 445-462.

Simonton, D. K. (2003). Qualitative and quantitative analyses of historical data. *Annual Review of Psychology, 54*(1), 617-640.

Sipos, Y., Battisti, B., & Grimm, K. (2008). Achieving transformative sustainability learning: engaging head, hands and heart. *International Journal of Sustainability in Higher Education, 9*(1), 68-86.

Socio-economic Trends - Education. (1996).   Retrieved from https://web.archive.org/web/20080526152536/http://www.dec-ced.gc.ca/Complements/Publications/AutresPublications-EN/tocen/css/tocen_15.htm

Soleri, P. (1984). *Arcosanti, an Urban Laboratory?*

Sondeijker, S., Geurts, J., Rotmans, J., & Tukker, A. (2006). Imagining sustainability: the added value of transition scenarios in transition management. *foresight, 8*(5), 15-30.

Sreedharan, E. (2007). *A manual of historical research methodology*

Staley, D. J. (2002). A History of the Future. *History and Theory, 41*(4), 72-89.

Steele, R. D. (2012). *The open-source everything manifesto: Transparency, truth, and trust*: North Atlantic Books.

Steinberg, P. F. (2015). Can We Generalize from Case Studies? *Global Environmental Politics.*

Stick to the finish. (1925, January 28, 1925). *The Marcellus Observer.*

Stickler, W. H. (1964). Experimental Colleges. *Tallahassee: Florida State University.*

Stillman, P. G. (1990). Recent Studies in the History of Utopian Thought: JSTOR.

Strauch, Y. (2002). An Emerging Campus Ecovillage Movement.   Retrieved from http://www.ulsf.org/pub_declaration_spotvol61.htm

Sustainability science a framework for the future. (2016).   Retrieved from http://www.unity.edu/about-unity/sustainability-science

Sydow, B. C. (2012). *Sustainability Design in Higher Education: Curriculum, Teaching Methods, and Program Integration*: ERIC.

Talen, E. (1999). Sense of community and neighbourhood form: An assessment of the social doctrine of new urbanism. *Urban Studies, 36*(8), 1361-1379.

Tappi, D. (2001). *The neo-marshallian industrial district. A study on Italian contributions to theory and evidence.* Paper presented at the Conferencia DRUID.

Taylor, J. (2014). *Boston Main Streets 2.0: Spreading Boston's Innovation Economy from the Innovation District to the Neighborhoods.* Tufts University.

Thompson Klein, J. (2004). Prospects for transdisciplinarity. *Futures, 36*(4), 515-526.

Tinsley, S., & George, H. (2006). Ecological footprint of the Findhorn foundation and community. *Sustainable Development Research Centre, Morray.*

Tischler, L. (2009). Ideo's David Kelley on "Design Thinking": David Kelley, founder of the design firm Ideo and the Stanford d. school, was leading a charmed existence. *17*, 2014.

Tolstoy, L., Pevear, R., & Volokhonsky, L. (1995). *What is art?* : Penguin UK.

Tornatzky, L. (2014). *Innovation U 2.0 Reinventing University Roles in a Knowledge Economy*   Retrieved from http://www.innovation-u.com

Tornatzky, L. G., Waugaman, P. G., & Gray, D. O. (2002). *Innovation U.: New university roles in a knowledge economy*: Southern Technology Council Research Triangle Park, NC.

Townsend, A., Soojung-Kim Pang, A., & Weddle, R. (2009). Future knowledge ecosystems: the next twenty years of technology-led economic development. *Institute for the Future, IFTF Report Number SR-12361*.

Trainer, T. (1998). Viewpoint: Towards a checklist for Ecovillage development. *Local Environment, 3*(1), 79-83.

USGBC. (2013). *LEED in motion: people and progress.* Retrieved from

Valdez, J. (2000). *The entrepreneurial ecosystem: toward a theory of new business formation.* Paper presented at the SMALL BUSINESS INSTITUTE DIRECTOR'S ASSOCIATION.

Van der Ryn, S. (2005). *Design for life: the architecture of Sim Van der Ryn*: Gibbs Smith.

Van der Ryn, S. (2012). Retrieved from http://www.greenplanetarchitects.com/en/project/details/berea-ecovillage

Van Evera, S. (1997). *Guide to methods for students of political science*: Cornell University Press.

van Oort, F. G., Burger, M., & Raspe, O. (2008). Inter-firm relations and economic clustering in the Dutch Randstad region. *The Economics of Regional Clusters*, 145-165.

Vincent, S., Bunn, S., & Stevens, S. (2012). Results from the 2012 census of US Four Year Colleges and Universities.

Vincent, S., Danielson, A., & Santos, B. R. (2015). Interdisciplinary Environmental and Sustainability Education and Research: Institutes and Centers at US Research Universities *Integrative Approaches to Sustainable Development at University Level* (pp. 275-292): Springer.

Vincent, S., Roberts, J. T., & Mulkey, S. (2015). Interdisciplinary environmental and sustainability education: islands of progress in a sea of dysfunction. *Journal of Environmental Studies and Sciences*, 1-7.

. Vision 2034. (2015): NC State.

Vollrath, C. J. (2012a). *After progress: The image of the future in the age of sustainability.* (3526933 Ph.D.), The University of Iowa, Ann Arbor.

ProQuest Dissertations & Theses Full Text database.

Vollrath, C. J. (2012b). After progress: the image of the future in the age of sustainability.

Von Krogh, G., & Von Hippel, E. (2006). The promise of research on open source software. *Management science, 52*(7), 975-983.

Wagar, W. W. (1998). Past and future. *American Behavioral Scientist, 42*(3), 365-371.

Wagner, F., Andreas, M., & Mende, S. (2012). Research in Community: Collaborating for a Culture of Sustainability. *Realizing Utopia*, 95.

Wagner, G. (1976). *The end of education*: AS Barnes.

Wagner, J. (1985). Success in intentional communities: The problem of evaluation. *Communal Societies, 5*, 89-100.

Wake Up & Dream project to host advocate of ecovillages across the world. (2012). Retrieved from http://spears.okstate.edu/news/2012/10/17/wake-up-dream-project-to-host-advocate-of-ecovillages-across-the-world/

Walker, L. (2005). *Ecovillage at Ithaca: pioneering a sustainable culture*: New Society Publishers.

Walker, L. (2012a). Coming of Age: 21 Years of Eco Village Planning and Living. *Communities*(156), 36-40.

Walker, L. (2012b). EcoVillage at Ithaca: Principles, Best Practices & Lessons Learned. *EcoVillage at Ithaca Center for Sustainability Education.*

Walmart. (2015). Sustainability Leaders. Retrieved from http://corporate.walmart.com/global-responsibility/environment-sustainability/sustainability-index-leaders-shop

Walsh, R. M. (2013). *The Origins of Vancouverism: A historical inquiry into the architecture and urban form of Vancouver, British Columbia.* University of Oregon.

Wang, Y., Shi, H., Sun, M., Huisingh, D., Hansson, L., & Wang, R. (2013). Moving towards an ecologically sound society? Starting from green universities and environmental higher education. *Journal of Cleaner Production, 61*, 1-5.

Ward, S. (2005). *The garden city: Past, present and future*: Routledge.

Way, T. (2016). The urban university's hybrid campus. *Journal of Landscape Architecture, 11*(1), 42-55.

Wessner, C. W. (2012). *Clustering for 21st Century Prosperity: Summary of a Symposium*: National Academies Press.

Whitney, M. K. (2014). *Voluntary University Sustainability Commitments: a Network in and of Transition.* Prescott College.

Wilde, O. (1950). *The Soul of Man Under Socialism, and Other Essays* (Vol. 204): Harper & Row.

Williams, J. (2005). Designing Neighbourhoods for Social Interaction: The Case of Cohousing. *Journal of Urban Design, 10*(2), 195-227.

Williams, J. (2008). Predicting an American future for cohousing. *Futures, 40*(3), 268-286.

Winkler, E. (2014). The Innovation Myth: Why you can't engineer creativity through innovation districts

. Retrieved from http://www.newrepublic.com/article/118815/innovation-districts-are-oversold-you-cant-engineer-creativity

Winthrop, H. (1962). Interdisciplinary Studies: Variations in Meaning, Objectives, and Accomplishments. *ADE Bulletin, 33*, 29-41.

Winthrop, H. (1963). Scientific, intentional communities can save democracy and religion. *Social Order, 13*, 21-31.

Wolcott, H. F. (1990). On seeking-and rejecting-validity in qualitative research. *Qualitative inquiry in education: The continuing debate*, 121-152.

Xiaojian, L. (2011). New industrial district and globalization: a literature review. *Progress In Geography, 16*(3), 16-23.

Xue, J. (2014). Is eco-village/urban village the future of a degrowth society? An urban planner's perspective. *Ecological Economics, 105*, 130-138.

Yu, J., & Jackson, R. (2011). Regional innovation clusters: A critical review. *Growth and Change, 42*(2), 111-124.

Zablocki, B. (1971). The Joyful Community. *Baltimore: Penguin Books.*

www.ingramcontent.com/pod-product-compliance
Lightning Source LLC
Chambersburg PA
CBHW081109170526
45165CB00008B/2378